T0134952

Springer Geography

The Springer Geography series seeks to publish a broad portfolio of scientific books, aiming at researchers, students, and everyone interested in geographical research.

The series includes peer-reviewed monographs, edited volumes, textbooks, and conference proceedings. It covers the major topics in geography and geographical sciences including, but not limited to; Economic Geography, Landscape and Urban Planning, Urban Geography, Physical Geography and Environmental Geography.

Springer Geography — now indexed in Scopus

Wuttichai Boonpook • Zhaohui Lin
Pakorn Meksangsouy • Parichat Wetchayont
Editors

Applied Geography and Geoinformatics for Sustainable Development

Proceedings of ICGGS 2022

 Springer

Editors
Wuttichai Boonpook
Department of Geography
Faculty of Social Science
Srinakharinwirot University
Bangkok, Thailand

Pakorn Meksangsouy
Department of Geography
Faculty of Social Science
Srinakharinwirot University
Bangkok, Thailand

Zhaohui Lin
International Center for Climate
and Environment Science (ICCES)
Institute of Atmospheric Physics
Chinese Academy of Sciences
Beijing, China

Parichat Wetchayont
Department of Geography
Faculty of Social Science
Srinakharinwirot University
Bangkok, Thailand

ISSN 2194-315X   ISSN 2194-3168   (electronic)
Springer Geography
ISBN 978-3-031-16219-0   ISBN 978-3-031-16217-6   (eBook)
https://doi.org/10.1007/978-3-031-16217-6

Contents

Flood Susceptibility Mapping Using a Frequency Ratio Model: A Case Study of Chai Nat Province, Thailand

Chanita Duangyiwa (iD) **and Pannee Cheewinsiriwat** (iD)

Abstract Flooding has become more prevalent in many regions of Southeast Asian countries in recent decades. Intense precipitation, settlement in low-lying areas, population growth, and rapid urbanization can enhance vulnerability to floods and lead to serious hazards. This study developed flood susceptibility mapping for the Chai Nat province of Thailand using flood-conditioning factors and the frequency ratio (FR) method. The flood inventory (2005–2017) was randomly separated into a training dataset for FR analysis and a testing dataset for model validation. Eleven flood-conditioning parameters, i.e., altitude, slope, curvature, the topographic wetness index, rainfall, distance to drainage, drainage density, soil drainage, land use, the normalized difference vegetation index, and road density, were considered for this study. While constructing the flood susceptibility index, the relative frequency and predictor rate were used to create the flooding probability for each factor class and the weight of each factor in the model. The values for the flood susceptibility index were classified into five categories and used to make a flood susceptibility map. The area under the curve (AUC) was used to validate the model prediction. The results indicate that the AUC values for the success and prediction rates are 74.2% and 75.1%, respectively.

Keywords Flood susceptibility · Frequency ratio · Predictor rate · GIS · Thailand

C. Duangyiwa (✉) · P. Cheewinsiriwat
Center of Excellence in Geography and Geoinformatics, Faculty of Arts, Chulalongkorn University, Bangkok, Thailand
e-mail: chanita.d@chula.ac.th

© The Author(s), under exclusive license to Springer Nature
Switzerland AG 2023
W. Boonpook et al. (eds.), *Applied Geography and Geoinformatics for Sustainable Development*, Springer Geography,
https://doi.org/10.1007/978-3-031-16217-6_1

1

1   Introduction

Flooding has become more frequent in recent years due to climate change. Most Southeast Asian countries, including Myanmar, Indonesia, Malaysia, Vietnam, and Thailand, have been affected by flooding in recent decades [1]. Intense precipitation is the most common cause of flooding, but socioeconomic factors, such as settlement in low-lying areas, rapid urbanization, and economic development, have increased flood risk exposure [2, 3]. Early human settlements were clustered near rivers and on low-lying terrain. Trends of preferential settlement and urbanization have resulted in a significant increase in the number of people living in the area. These expose areas to different types of floods, such as river overflow, surface water flow, coastal inundation, and flash flooding [4, 5].

Chai Nat is a province in the Chao Phraya River basin, the central part of Thailand. Most of the region is lowland, bordered by rivers and irrigation canals. In the past, flooding in Chai Nat has resulted from high precipitation in the upstream basin. Extreme weather events that involve variations in rainfall patterns and/or tropical storms could increase overall precipitation and runoff volume. In addition, the development of social and economic systems may affect the spatial patterns of flood risk. Population growth could influence changes in types of land use. The construction of new structures and infrastructure may limit floodplain drainage and increase its susceptibility to flooding [6]. Hence, the development of flood modeling to predict vulnerable areas could benefit flood adaptation and mitigation.

The frequency ratio (FR) is a bivariate statistical model that has been used to map flood susceptibility in several studies, in Malaysia, Korea, China, Iran, Papua New Guinea, Pakistan, and Thailand [7–15], among others. The FR approach investigated the ratio of the probability of flooding to the probability of nonflooding for given characteristics to assess the influences of classes of each conditioning factor on flood occurrence [10]. Various flood-hazard-influencing factors, such as elevation, slope, aspect, curvature, land use, rainfall, drainage density, soil drainage, topographic wetness index (TWI), and distance to river, must be specified to develop FR model [9, 11, 12]. The success and prediction rates in preliminary studies using the FR model varied between 64–96% and 60–97%, respectively [10–15].

The FR model has been used in a few studies to assess flood risk in Thailand. Anuchan and Iamchuen [14] employed the FR approach to map flood susceptibility in southern Thailand's Songkhla Lake basin. In the study, only six contributing parameters were used: elevation, slope, drainage density, road density, soil drainage, and land use. Seejata et al. [15] used the FR process to create a flood hazard map in Sukhothai province, taking into account eight parameters: rainfall, elevation, slope, soil drainage, land use, drainage density, road density, and distance to the river. The validation results from these studies demonstrated that the FR approach was successful for flood risk mapping in both areas. In Chai Nat, there is currently no clear evidence for influencing factors or their relationship to flood occurrences. Therefore, this study simulates flood susceptibility mapping in Chai Nat using the geospatial FR technique. The 11 significant elements, i.e., altitude, slope, curvature, TWI, rainfall, distance to drainage, drainage density, soil drainage, land use, normalized difference vegetation index (NDVI), and road density, were used in this model.

2 Study Area

Chai Nat is located in the central region of Thailand, covering an area of 2470 km² (Fig. 1). Floodplains characterize the topography of this area, with elevations lower than 296 m above sea level. The study area is situated in the Chao Phraya River basin, the largest in Thailand, covering approximately 30% of the country's total area [16, 17]. The basin includes two major river systems: the Chao Phraya and the Tha Chin. The Chao Phraya River begins at the confluence of the Ping, Wang, Yom, and Nan Rivers, all of which originate in northern Thailand. The Tha Chin River is a tributary of the Chao Phraya River, which branches off the main river in Chai Nat and flows to the Gulf of Thailand [18].

The average annual precipitation is approximately 1000 mm. The wettest month is September, and the area has an average of 20 rainy days and 240 mm precipitation per month. The average temperature is about 27.8 °C. The temperature and precipitation records for the period 1970–2020 indicate that the greatest maximum temperature was 41.8 °C in April 2016 and the highest maximum precipitation was 130.1 mm in April 2000 [19].

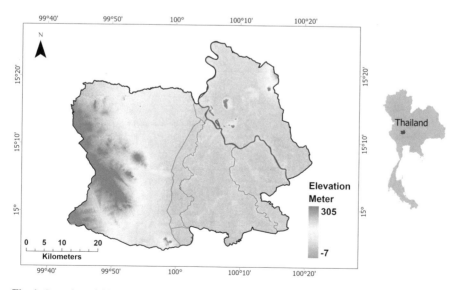

Fig. 1 Location of Chai Nat, Thailand

In 1957, the Chao Phraya Dam was built on the Chao Phraya River to enable irrigation and electricity generation [20]. This dam is situated in the Sapphaya district of Chai Nat and has a maximum discharge capacity of 3200 m²/s [21]. Flooding could be reduced by temporarily storing runoff in a reservoir and gradually releasing it over time to limit flow and flooding downstream. Although the dam was built to mitigate the effects of flooding, those who live above its dike are nevertheless affected by flooding. Increasing water volume at the confluence of tributaries causes the discharge to increase as it flows downstream. Due to the lack of discharge capacity in the river's downstream reach, flooding caused by overtopping is frequent [22].

3   Methodology

This study used the FR method to generate a flood susceptibility map (Fig. 2). First, the required data for a flood inventory map and flood-conditioning factors was collected (Table 1). A map of the flood inventory (2005–2017) was generated from historical data. The relationship between the locations of flood occurrences and flood-conditioning factors was then analyzed using FR algorithms. The predictor rate (PR) or weights for each flood-conditioning factor were then determined using each factor class's relative frequency (RF) values. For the simulation of flood susceptibility mapping, the computed RF and PR values were applied to each class within the different factors. The accuracy of flood prediction was examined using the area under the curve (AUC) values.

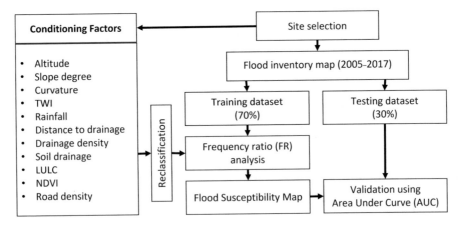

Fig. 2 Methodology flowchart for flood susceptibility mapping

Table 1 Dataset and data source used for flood susceptibility assessment

Dataset	Data source and data description	Scale and resolution
Flood extent	GISTDA (2005–2017)	
Altitude	ALOS PALSAR DEM	12.5 m
Slope	Derived from DEM	12.5 m
Curvature	Derived from DEM	12.5 m
TWI	Derived from DEM	12.5 m
Rainfall	Thai Meteorological Department (1987–2017)	
Distance to drainage	Department of Water Resources (2018)	1: 50,000
Drainage density	Department of Water Resources (2018)	1: 50,000
Soil drainage	Land Development Department (2017)	1: 25,000
LULC	Land Development Department (2017)	1: 25,000
NDVI	Sentinel-2 MSI imagery (December 23, 2017)	10 m
Road density	Ministry of Transport (2012)	1: 20,000

3.1 Flood Inventory Map

The flood inventory map displays the spatial distribution of flood hazards in the study area [11]. To predict the risk of flooding, it is necessary to investigate the relationship between flood occurrence and the conditioning parameters [12, 23]. The flood inventory map was generated using data obtained from the national database of the Geo-Informatics and Space Technology Development Agency (GISTDA), based on the inundation extent of flood events that occurred across Chai Nat from 2005 to 2017. The 1979 flooded locations were generated every 700 m within the flooded areas and then randomly separated into 70% training dataset for mapping flood vulnerability and 30% testing dataset for evaluating the model's predictions.

3.2 Flood-Conditioning Factors

Flood-conditioning factors must be specified in relation to the characteristics of a geographical location to generate a flood susceptibility map [11, 1224 24]. This study employs 11 flood-conditioning factors: altitude, slope, curvature, TWI, soil drainage, land use and land cover (LULC), rainfall, NDVI, distance to drainage, drainage density, and road density (Table 1). Each factor, except for curvature, soil drainage, and land use, was resampled and converted to a 12.5 × 12.5 m pixel size. The quantile classification method was used to classify the factors into ten classes to apply the FR technique.

The altitude map was acquired from the 12.5 m resolution ALOS DEM produced by the Alaska Satellite Facility Distributed Active Archive Data Center. The slope degree and curvature were computed from elevation of the DEM. The curvature was categorized into three classes: a negative curvature indicates concavity, a zero curvature indicates a flat area, and a positive curvature depicts convexity. The yearly average rainfall data was compiled from ten Thai Meteorological Department rainfall stations over a period of 30 years (1987–2017) using the inverse distance weighting technique to interpolate the rainfall in the study area.

The TWI is a physical depiction of flood inundation regions. It can determine the amount of geographic variation for runoff generation, as it indicates the distribution of moisture and groundwater flow at different locations [25, 26]. The TWI is calculated using Eq. (1) as follows:

$$TWI = \ln\left(\frac{A_s}{\tan \beta}\right) \tag{1}$$

where A_s is the upslope contributing area and β is the topographic slope measured in degree at the point [24].

The proximity to rivers increases the risk of flooding, and an intensive drainage density is responsible for increased surface runoff [25–27]. Moreover, road density functions as an obstacle, along with drainage density networks, and it could cause waterlogging [14, 15]. The impact of drainage channels on flood vulnerability was investigated by determining the proximity to rivers by calculating the Euclidean distance. Drainage density and road density are generated using the line density calculation.

Surface runoff was influenced by soil drainage, as there is less surface runoff from well-drained soils than poorly drained ones [9, 14]. Soil drainages were grouped into seven categories: well drained, moderately well drained, somewhat poorly drained, poorly drained, very poorly drained, slope complex, and water.

Land use influences flooding, as it affects hydrological processes, such as permeability and runoff [8, 26]. Urban environments enhance runoff due to the extensive impermeable soil. Uncultivated farmland increases runoff, as there is no vegetation to control the water flow on the soil surface; hence, there is a risk of flooding in such areas [13]. Land use was reclassified into seven categories: agriculture, forest, grass, shrubland, bare, urban, and water.

The NDVI is an indicator of vegetation used to monitor its characteristics on the earth's surface using satellite images. The index admits of values between −1 and +1. The NDVI reflects the health of vegetation health and can be used to indicate resistance to flooding [12, 25, 28]. The NDVI index was derived from Sentinel-2 MSI imagery on December 23, 2017 and was calculated by Eq. (2) as follows:

$$NDVI = \frac{NIR - RED}{NIR + RED} \tag{2}$$

where NIR is the near-infrared band and RED is the red band.

3.3 Frequency Ratio Method

The FR approach measures the association between flood occurrences and conditioning factors. The level of association is indicated by FR values [29]. A greater FR value suggests a greater probability of flooding. FR values greater than 1 show stronger correlations, whereas values less than 1 suggest weaker relationships with flood occurrences [10, 12]. FR values are determined using Eq. (3).

$$FR = \frac{\text{Number of flood in the factor class / Total flood locations}}{\text{Number of pixels in the factor class / Total pixels}} \quad (3)$$

The FR was then normalized as RF (RF) or weight of each class in a range of probability values from 0 to 1 in Eq. (4).

$$RF = \frac{FR \text{ of the Factor class}}{\text{Sum } FR \text{ of each factor class}} \quad (4)$$

After normalization, the RF values were then used in Eq. (5) to calculate a PR or weight of each factor by assessing each flood causative factor using the training dataset [13, 30, 31]:

$$PR = \frac{RF_{\text{max}} - RF_{\text{min}}}{\min \left(RF_{\text{max}} - RF_{\text{min}} \right)} \quad (5)$$

where RF_{max} and RF_{min} are the maximum and minimum RF of each factor, respectively, and $\text{Min}(RF_{\text{max}} - RF_{\text{min}})$ is the lowest difference of all factors.

The flood susceptibility index (FSI) was then calculated using Eq. (6) by adding the results for multiplying PR of each factor by the RF of each class in that factor. The flood susceptibility map was generated to illustrate the FSI of each pixel in the study area:

$$FSI = \sum_{i=1}^{n} PR_i \times RF_i \quad (6)$$

where PR_i is the weight of each flood-conditioning factor, RF is the weight of each factor class of the flood-conditioning factor, and n is the number of factors.

3.4 Model Validation

The receiver operating characteristics curve was used for model validation to evaluate the performance and accuracy of a flood susceptibility map [12, 28]. The AUC was used to validate the model prediction. The validation procedure was performed by comparing the existing flood locations to the generated flood susceptibility map. The success and prediction rates of the AUC were calculated using 70% training data and 30% testing data, respectively.

The success rate indicates how well the model fits the training dataset [32]. The prediction rate indicates the model's prediction accuracy. The prediction rates were calculated by comparing flood susceptibility maps to the flood testing dataset. Models with higher AUC values for success and prediction are considered more accurate and reliable [7].

4   Results and Discussion

In this study, 1979 points from the flood inventory of 2827 flooded points were randomly selected as the training dataset for the FR model, while the remaining 848 points were used as the testing dataset for the model validation (Fig. 3). The FR values and PRs were calculated for the 11 conditioning factors (Fig. 4), including altitude, slope, curvature, TWI, rainfall, distance to drainage, drainage density, soil drainage, LULC, NDVI, and road density, based on their relationship to the flooding.

Table 2 shows the correlation between flood-conditioning factors and flood occurrences, with the highest FR values of each factor highlighted in bold.

The spatial pattern of the altitude map was constructed after reclassifying (Fig. 4a). The altitude below 20 m is the most prevalent class, accounting for 47% of the total area but 87% of historical flooded locations. The altitude class between −7.37 and 12.2 m has the highest FR value, of 2.09, while at classes greater than 25.66 m, the FR values are less than 0.18. This suggests that low-elevation areas are more vulnerable to flooding.

Most of the study area has a low slope gradient (<5°) (Fig. 4b). Approximately 95% of flooded locations occurred in areas with slopes less than 6.13°. The FR values were greatest for slopes less than 0.27°, with a value of 1.20. Gentle slopes have a stronger correlation with flooding than the steep slope.

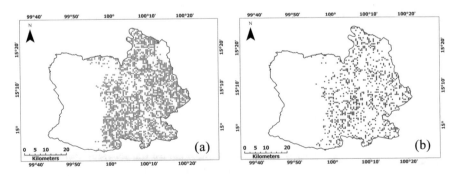

Fig. 3 The flood inventory map is divided into (**a**) training and (**b**) testing datasets

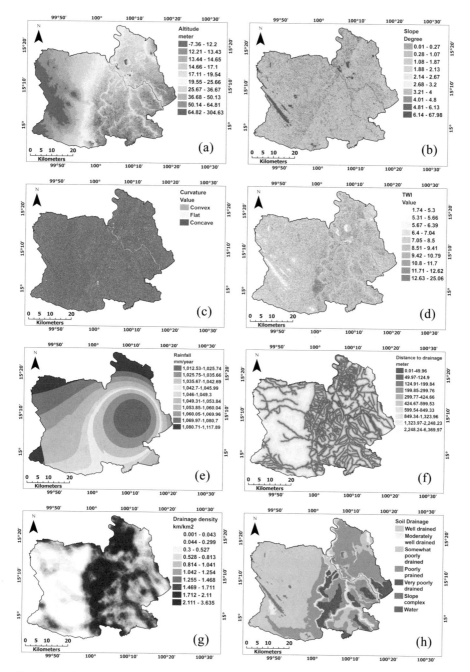

Fig. 4 Flood-conditioning factors in the study area: (**a**) altitude, (**b**) slope, (**c**) curvature, (**d**) TWI, (**e**) rainfall, (**f**) distance to drainage, (**g**) drainage density, (**h**) soil drainage, (**i**) LULC, (**j**) NDVI, and (**k**) road density

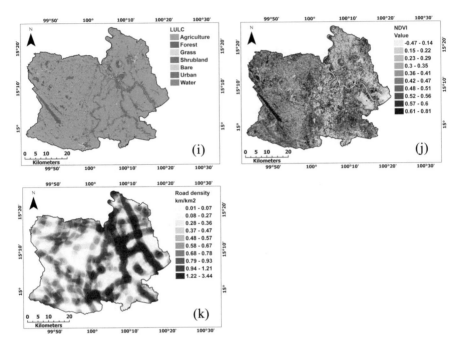

Fig. 1.4 (continued)

The results show that flat curvature with an FR value of 1.34 had the highest FR. This indicated that flat curvature areas have the highest probability of flooding, which echoes the results of Yarinyan et al. [26] and Bui et al. [28]. This is because flat areas may retain flood flows for an extended period [8].

The TWI values are higher in the eastern, northeastern, and southeastern parts of the study area and lower in steep slope zones (Fig. 4d). The results show that a higher TWI indicates a higher probability of floods in that area. The TWI class of 11.7–12.62 has the highest FR value of 1.38.

The rainfall class of 1035.66–1042.69 mm has the highest FR value of 1.73. Approximately 66% of flooded locations occur in areas with minimal rainfall. It is worth noting that this study area is defined as a floodplain, and floods occur in low-land areas to the east and southeast, where runoff from the upstream rivers in the north has accumulated in a downstream river.

The result shows that most flooded locations occurred within 0–850 m from a river. The distance between 199.84 and 299.76 m from the river correlates with floods, with the highest FR value of 1.48.

The drainage density region between 1.04 and 1.25 km/km^2 has the highest FR value, 1.68. Areas with drainage density higher than 1.25 km/km^2 have almost the same level of probability to flooding.

Sixty percent of this area is poorly drained soil, located on both sides of the Chao Phraya River (Fig. 4h). Most floods occurred in very poorly drained and poorly

Table 2 Calculation results for frequency ratio and relative frequency

Parameter	Class	Pixels in domain	% Domain	No of floods	% Floods	FR	RF
Altitude (meter)	−7.37–12.2	1,003,197	6.25	258	13.05	**2.09**	0.20
	12.2–13.43	863,889	5.38	202	10.22	1.90	0.18
	13.43–14.65	2,182,421	13.60	540	27.31	2.01	0.20
	14.65–17.1	1,959,290	12.21	436	22.05	1.81	0.18
	17.1–19.54	1,495,110	9.32	278	14.06	1.51	0.15
	19.54–25.66	2,376,078	14.81	217	10.98	0.74	0.07
	25.66–36.67	1,801,929	11.23	39	1.97	0.18	0.02
	36.67–50.13	1,463,358	9.12	0	0.00	0.00	0.00
	50.13–64.81	1,435,505	8.95	5	0.25	0.03	0.003
	64.81–304.63	1,465,190	9.13	2	0.10	0.01	0.001
Slope degree	0–0.27	1,114,600	6.96	165	8.35	**1.20**	0.12
	0.27–1.07	1,591,016	9.93	234	11.85	1.19	0.12
	1.07–1.87	2,177,634	13.59	287	14.53	1.07	0.11
	1.87–2.13	1,143,302	7.14	147	7.44	1.04	0.10
	2.13–2.67	2,885,291	18.01	360	18.23	1.01	0.10
	2.67–3.2	1,638,842	10.23	185	9.37	0.92	0.09
	3.2–4	1,952,780	12.19	203	10.28	0.84	0.08
	4–4.8	1,507,093	9.41	181	9.16	0.97	0.10
	4.8–6.13	1,033,095	6.45	118	5.97	0.93	0.09
	6.13–67.98	980,153	6.12	95	4.81	0.79	0.08
Curvature	Convex	7,476,773	46.60	902	45.62	0.98	0.30
	Flat	1,115,662	6.95	184	9.31	**1.34**	0.41
	Concave	7,453,532	46.45	891	45.07	0.97	0.30
TWI	1.73–5.3	1,193,112	7.44	90	4.56	0.61	0.06
	5.3–5.66	1,587,045	9.89	171	8.66	0.88	0.09
	5.66–6.39	1,890,989	11.78	190	9.62	0.82	0.08
	6.39–7.04	1,932,048	12.04	220	11.14	0.93	0.09
	7.04–8.5	1,759,467	10.96	173	8.76	0.80	0.08
	8.5–9.41	660,296	4.11	73	3.70	0.90	0.09
	9.41–10.79	2,241,615	13.97	284	14.38	1.03	0.10
	10.79–11.7	1,628,008	10.15	251	12.71	1.25	0.13
	11.7–12.62	1,602,918	9.99	272	13.77	**1.38**	0.14
	12.62–25.06	1,528,308	9.52	251	12.71	1.33	0.13
Rainfall (mm)	1012.52–1025.74	1,585,660	9.88	314	15.88	1.61	0.16
	1025.74–1035.66	1,634,176	10.18	334	16.89	1.66	0.17
	1035.66–1042.69	1,665,764	10.38	356	18.01	**1.73**	0.17
	1042.69–1045.99	1,499,737	9.35	302	15.28	1.63	0.16
	1045.99–1049.3	1,667,941	10.39	180	9.10	0.88	0.09
	1049.3–1053.84	1,679,751	10.47	101	5.11	0.49	0.05
	1053.84–1060.04	1,668,546	10.40	111	5.61	0.54	0.05
	1060.04–1069.96	1,586,454	9.89	92	4.65	0.47	0.05
	1069.96–1080.7	1,545,991	9.63	64	3.24	0.34	0.03
	1080.7–1117.89	1,512,192	9.42	123	6.22	0.66	0.07

(continued)

Table 2 (continued)

Parameter	Class	Pixels in domain	% Domain	No of floods	% Floods	FR	RF
Distance to drainage (m)	0–49.96	1,379,437	8.60	184	9.31	1.08	0.11
	49.96–124.9	1,963,012	12.23	295	14.92	1.22	0.13
	124.9–199.84	1,634,979	10.19	256	12.95	1.27	0.13
	199.84–299.76	1,770,322	11.03	322	16.29	**1.48**	0.15
	299.76–424.66	1,709,646	10.65	282	14.26	1.34	0.14
	424.66–599.53	1,664,851	10.38	271	13.71	1.32	0.14
	599.53–849.33	1,522,091	9.49	196	9.91	1.05	0.11
	849.33–1323.96	1,539,128	9.59	141	7.13	0.74	0.08
	1323.96–2248.23	1,441,415	8.98	25	1.26	0.14	0.01
	2248.23–6369.97	1,421,331	8.86	5	0.25	0.03	0.003
Drainage density (km/km^2)	0–0.04	1,478,231	9.21	6	0.30	0.03	0.003
	0.04–0.3	1,721,514	10.73	22	1.11	0.10	0.01
	0.3–0.53	1,636,256	10.20	58	2.93	0.29	0.03
	0.53–0.81	1,591,087	9.92	196	9.91	1.00	0.10
	0.81–1.04	1,625,320	10.13	223	11.28	1.11	0.11
	1.04–1.25	1,625,423	10.13	337	17.05	**1.68**	0.17
	1.25–1.47	1,600,119	9.97	298	15.07	1.51	0.15
	1.47–1.71	1,583,090	9.87	251	12.70	1.29	0.13
	1.71–2.11	1,629,128	10.15	287	14.52	1.43	0.14
	2.11–3.63	1,556,044	9.70	299	15.12	1.56	0.16
Soil drainage	Well drained	4,908,665	30.59	44	2.23	0.07	0.11
	Moderately well drained	988,111	6.16	80	4.05	0.66	0.01
	Somewhat poorly drained	3,166,346	19.73	294	14.87	0.75	0.31
	Poorly drained	5,050,269	31.47	1159	58.62	1.86	0.13
	Very poorly drained	1,551,897	9.67	390	19.73	**2.04**	0.02
	Slope complex	288,497	1.80	5	0.25	0.14	0.07
	Water	92,427	0.58	5	0.25	0.44	0.34
LULC	Agriculture	12,917,535	80.50	1777	89.88	**1.12**	0.28
	Forest	472,632	2.95	9	0.46	0.15	0.04
	Grass	58,207	0.36	3	0.15	0.42	0.11
	Shrubland	124,582	0.78	2	0.10	0.13	0.03
	Bare	103,233	0.64	11	0.56	0.86	0.22
	Urban	1,659,917	10.34	111	5.61	0.54	0.14
	Water	710,106	4.43	64	3.24	0.73	0.18

(continued)

Table 2 (continued)

Parameter	Class	Pixels in domain	% Domain	No of floods	% Floods	FR	RF
NDVI	−0.48–0.14	10,030	0.06	0	0.00	0.00	0.00
	0.14–0.22	89,018	0.55	7	0.35	0.64	0.06
	0.22–0.29	67,975	0.42	10	0.51	1.19	0.12
	0.29–0.35	212,326	1.32	55	2.78	**2.10**	0.21
	0.35–0.41	1,566,506	9.76	326	16.49	1.69	0.17
	0.41–0.46	3,005,588	18.73	410	20.74	1.11	0.11
	0.46–0.51	3,577,578	22.30	383	19.37	0.87	0.09
	0.51–0.56	4,372,579	27.25	464	23.47	0.86	0.08
	0.56–0.6	2,984,010	18.60	305	15.43	0.83	0.08
	0.6–0.78	160,602	1.00	17	0.86	0.86	0.08
Road density (km/km²)	0–0.07	1,553,104	9.68	209	10.57	1.09	0.11
	0.07–0.27	1,502,292	9.36	179	9.05	0.97	0.10
	0.27–0.36	1,643,072	10.24	179	9.05	0.88	0.09
	0.36–0.47	1,618,246	10.08	188	9.51	0.94	0.09
	0.47–0.57	1,670,475	10.41	171	8.65	0.83	0.08
	0.57–0.67	1,765,991	11.01	188	9.51	0.86	0.09
	0.67–0.78	1,567,000	9.77	136	6.88	0.70	0.07
	0.78–0.93	1,613,957	10.06	215	10.88	1.08	0.11
	0.93–1.21	1,612,101	10.05	266	13.45	**1.34**	0.13
	1.21–3.44	1,499,974	9.35	246	12.44	1.33	0.13

drained categories, with FR values of 2.04 and 1.86. The results demonstrate that when soil drainage capacity decreases, FR values increase.

For LULC, the results show that agricultural regions have the greatest FR value, 1.12, consistent with Ullah and Zhang's findings [13] and those of Yariyan et al. [26]. Shrubland and forest have the lowest FR values, 0.13 and 0.15, respectively.

NDVI values between 0.29 and 0.35 show the highest correlation with flooding, with an FR value of 2.1. A link was found between flood occurrences and NDVI values between 0.29 and 0.60, with FR values decreasing as NDVI values increased.

Road density between 0.93 and 1.21 km/km² has the highest FR value of 1.34. On the contrary, the impact of road density on flooding is difficult to determine, as the trend from the road density classes is quite uncertain.

The PRs of each factor were calculated using the RF values in Table 2 and Eq. (5). Table 3 shows that soil drainage has the highest PR, with a value of 6.6, followed by LULC, altitude, and NDVI with PR values of 5, 4, and 4, respectively. This means that these variables significantly impact flooding in the study area. Drainage density, distance to drainage, rainfall, and curvature all have a medium impact on flood occurrence. TWI, road density, and slope contribute less to flooding than the other factors investigated in this study.

Table 3 Predictor rates or weights of the conditioning factors

Parameter	RF_{Max}	RF_{Min}	$RF_{Max} - RF_{min}$	Min $(RF_{Max} - RF_{min})$	PR
Altitude	0.20	0.00	0.20	0.05	4.00
Slope	0.12	0.07	0.05	0.05	1.00
Curvature	0.40	0.29	0.11	0.05	2.20
TWI	0.13	0.06	0.07	0.05	1.40
Rainfall	0.17	0.03	0.14	0.05	2.80
Distance to drainage	0.15	0.003	0.147	0.05	3.00
Drainage density	0.16	0.003	0.157	0.05	3.20
Soil drainage	0.34	0.01	0.33	0.05	6.60
LULC	0.28	0.03	0.25	0.05	5.00
NDVI	0.20	0.00	0.20	0.05	4.00
Road density	0.13	0.07	0.06	0.05	1.20

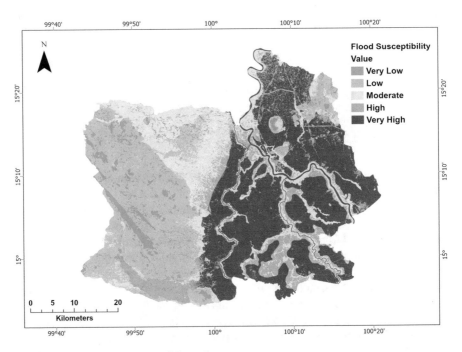

Fig. 5 Flood susceptibility map of the study area

The FSI values for each pixel were calculated using the factors' RF values and PR weights, applying Eq. (6). The computed flood susceptibility map is displayed in Fig. 5. FSI values in this study ranged from 149.6 to 780. The FSI index map was divided into five categories using the natural breaks method: extremely low (149.6–261.1), low (261.1–377.7), moderate (377.7–496.8), high (496.8–618.5), and very high (618.5–796), as shown in Table 4. Most high- to very-high-risk areas are in the east, with very poorly drained soil, low altitude, low NDVI, and higher

Table 4 Flood hazard susceptibility classification in Chai Nat, Thailand

Susceptibility	Area (km^2)	Ratio (%)
Very low	103.81	4.15
Low	717.24	28.65
Moderate	455.96	18.21
High	373.30	14.91
Very high	853.41	34.91

drainage densities. The classification of the study is as follows: 48% has high to very high susceptibility to flooding, 18% has medium susceptibility, and around 32% of the total area has low and very low susceptibility.

The AUC was used to validate the results of this study. The success rate and the prediction rate curve were calculated. Having a higher AUC indicates that the model is better at prediction. From this result, the AUC values for success and prediction rates are 74.2% and 75.1%, respectively.

5 Conclusion

This study used the FR approach to simulate flood susceptibility mapping in Chai Nat, Thailand. A flood inventory map was generated from historical data and then separated into training and testing datasets. FSI values were calculated using the training data and the association between the conditioning parameters and flood occurrences in the study area. Altitude, slope, curvature, TWI, soil drainage, LULC, rainfall, NDVI, distance to drainage, drainage density, and road density are all flood-conditioning factors. The susceptibility mapping derived five flood-susceptible classifications for the study site. According to the findings, around 35% of the study area is very susceptible to flooding. Floods were more frequent in agricultural land with very poorly drained soil, altitudes less than 26 m, slopes less than 6°, and lower NDVI values but higher TWI and drainage density values. The FR model performed well, with a 74.2% success rate and a 75.1% prediction rate. The challenge for future work will be to integrate models such as logistic regression and weights of evidence to develop a greater understanding of flood disaster mitigation. The findings presented here can serve as a starting point for further flood assessment and could provide a better knowledge of flood vulnerability in the study area.

Acknowledgments The authors would like to thank the Geo-Informatics and Space Technology Development Agency (GISTDA), Thai Meteorological Department, Department of Water Resources, Land Development Department, and the Ministry of Transport for providing the data employed in this research. We appreciate the Office of Research Affairs, Chulalongkorn University for supporting the Geography and Geoinformation Research Unit in conducting this study.

References

1. Syvitski, J.P.M., Kettner, A.J., Overeem, I., Hutton, E.W.H., Hannon, M.T., Brakenridge, G.R., Day, J., Vörösmarty, C., Saito, Y., Giosan, L., Nicholls, R.J.: Sinking deltas due to human activities. Nat. Geosci. **2**(10), 681–686 (2009)
2. Hirabayashi, Y., Mahendran, R., Koirala, S., Konoshima, L., Yamazaki, D., Watanabe, S., Kim, H., Kanae, S.: Global flood risk under climate change. Nat. Clim. Chang. **3**(9), 816–821 (2013)
3. IPCC: The Physical Science Basis. Cambridge University Press (2013)
4. Nicholls, R.J., Hanson, S., Herweijer, C., Patmore, N., Hallegatte, S., Corfee-Morlot, J., Château, J., Muir-Wood, R.: Ranking Port Cities with High Exposure and Vulnerability to Climate Extremes: Exposure Estimates. OECD Publishing (2008)
5. Nicholls, R.J., Cazenave, A.: Sea-level rise and its impact on coastal zones. Science. **328**(5985), 1517–1520 (2010)
6. Tehrany, M.S., Kumar, L., Shabani, F.: A novel GIS-based ensemble technique for flood susceptibility mapping using evidential belief function and support vector machine: Brisbane, Australia. PeerJ. **7**, 1–32 (2019)
7. Tehrany, M.S., Pradhan, B., Jebur, M.N.: Spatial prediction of flood susceptible areas using rule based decision tree (DT) and a novel ensemble bivariate and multivariate statistical models in GIS. J. Hydrol. **504**, 69–79 (2013)
8. Rahmati, O., Pourghasemi, H.R., Zeinivand, H.: Flood susceptibility mapping using frequency ratio and weights-of-evidence models in the Golastan Province, Iran. Geocarto Int. **31**(1), 42–70 (2016)
9. Samanta, S., Pal, D.K., Palsamanta, B.: Flood susceptibility analysis through remote sensing, GIS and frequency ratio model. Appl Water Sci. **8**(2), 1–4 (2018)
10. Lee, M., Kang, J., Jeon, S.: Application of frequency ratio model and validation for predictive flooded area susceptibility mapping using GIS. In: 2012 IEEE International Geoscience and Remote Sensing Symposium, pp. 895–898 (2012)
11. Cao, C., Xu, P., Wang, Y., Chen, J., Zheng, L., Niu, C.: Flash flood hazard susceptibility mapping using frequency ratio and statistical index methods in coalmine subsidence areas. Sustainability. **8**(9), 948 (2016)
12. Khosravi, K., Nohani, E., Maroufinia, E., Pourghasemi, H.R.: A GIS-based flood susceptibility assessment and its mapping in Iran: a comparison between frequency ratio and weights-of-evidence bivariate statistical models with multi-criteria decision-making technique. Nat. Hazards. **83**(2), 947–987 (2016)
13. Ullah, K., Zhang, J.: GIS-based flood hazard mapping using relative frequency ratio method: a case study of Panjkora River Basin, eastern Hindu Kush, Pakistan. PLoS One. **15**(3), e0229153 (2020)
14. Anucharn, T., Iamchuen, N.: Flood susceptibility map based on frequency ratio method at Songkhla Lake Basin in the Southern of Thailand. Burapha. Sci. J. **22**(3), 106–122 (2017)
15. Seejata, K., Yodying, A., Chatsudarat, S., Chidburee, P., Mahavik, N., Kongmuang, C., Tantanee, S.: Assessment of flood hazard using geospatial data and frequency ratio model in Sukhothai Province, Thailand. In: Proceeding of Asian Conference on Remote Sensing (2019)
16. Mikhailov, V.N., Nikitina, O.I.: Hydrological and morphological processes in the Chao Phraya Mouth Area (Thailand) and their anthropogenic changes. Water Resour. **36**(6), 613–624 (2009)
17. Rakwatin, P., Sansena, T., Marjang, N., Rungsipanich, A.: Using multi-temporal remote-sensing data to estimate 2011 flood area and volume over Chao Phraya River basin, Thailand. Remote Sens. Lett. **4**(3), 243–250 (2013)
18. Vongvisessomjai, S.: Chao Phraya Delta: Paddy field irrigation area in tidal deposit. In: Seminar on Irrigation Technologies for Sustainable Agricultural (2006)
19. Thai. Meteorological Department: http://climate.tmd.go.th/data/province/กลาง/ภูมิอากาศจังหวาท.pdf; http://climate.tmd.go.th/data/province/กลาง/ภูมิอากาศจังหวาท.pdf. Last accessed 12 Feb 2022
20. Haruyama, S.: Geomorphology of the central plain of Thailand and its relationship with recent flood conditions. GeoJournal. **31**(4), 327–334 (1993)

21. Electricity Generating Authority of Thailand: https://www.egat.co.th/home/chao-phraya-rohpp. Last accessed 12 Feb 2022
22. Komori, D., Nakamura, S., Kiguchi, M., Nishijima, A., Yamazaki, D., Suzuki, S., Kawasaki, A., Oki, K., Oki, T.: Characteristics of the 2011 Chao Phraya River flood in Central Thailand. Hydrol. Res. Lett. **6**, 41–46 (2012)
23. Gudiyangada, N.T., Tavakkoli, P.S., Gholamnia, K., Ghorbanzadeh, O., Rahmati, O., Blaschke, T.: Flood susceptibility mapping with machine learning, multi-criteria decision analysis and ensemble using Dempster Shafer Theory. J. Hydrol. **590**, 125275 (2020)
24. Shafapour, T.M., Shabani, F., Neamah, J.M., Hong, H., Chen, W., Xie, X.: GIS-based spatial prediction of flood prone areas using standalone frequency ratio, logistic regression, weight of evidence and their ensemble techniques. Geomat. Nat. Haz. Risk. **8**(2), 1538–1561 (2017)
25. Khoirunisa, N., Ku, C.Y., Liu, C.Y.: A GIS-based artificial neural network model for flood susceptibility assessment. Int. J. Environ. Res. Public Health. **18**(3), 1072 (2021)
26. Yariyan, P., Avand, M., Abbaspour, R.A., Torabi, H.A., Costache, R., Ghorbanzadeh, O., Janizadeh, S., Blaschke, T.: Flood susceptibility mapping using an improved analytic network process with statistical models. Geomat. Nat. Haz. Risk. **11**(1), 2282–2314 (2020)
27. Wu, Z., Shen, Y., Wang, H., Wu, M.: Assessing urban flood disaster risk using Bayesian network model and GIS applications. Geomat. Nat. Haz. Risk. **10**(1), 2163–2184 (2019)
28. Tien, B.D., Khosravi, K., Shahabi, H., Daggupati, P., Adamowski, J.F.M., Melesse, A., Thai Pham, B., Pourghasemi, H.R., Mahmoudi, M., Bahrami, S., Pradhan, B., Shirzadi, A., Chapi, K., Lee, S.: Flood spatial modeling in northern Iran using remote sensing and GIS: a comparison between evidential belief functions and its ensemble with a multivariate logistic regression model. Remote Sens. **11**(13), 1589 (2019)
29. Liuzzo, L., Sammartano, V., Freni, G.: Comparison between different distributed methods for flood susceptibility mapping. Water Resour. Manag. **33**(9), 3155–3173 (2019)
30. Na, T., Kawamura, Y., Kang, S.-S., Utsuki, S.: Hazard mapping of ground subsidence in east area of Sapporo using frequency ratio model and GIS. Geomat. Nat. Haz. Risk. **12**(1), 347–362 (2021)
31. Acharya, T.D., Lee, D.H.: Landslide susceptibility mapping using relative frequency and predictor rate along Araniko highway. KSCE J. Civ. Eng. **23**(2), 763–776 (2019)
32. Tehrany, M.S., Pradhan, B., Jebur, M.N.: Flood susceptibility analysis and its verification using a novel ensemble support vector machine and frequency ratio method. Stoch. Env. Res. Risk A. **29**(4), 1149–1165 (2015)

Influence of Hydrosphere Material Knowledge on the Attitude of High School Students in Conducting Water Conservation in Brebes Regency, Indonesia

Ristiani, Dede Rohmat, and Iwan Setiawan

Abstract Geography as an applied science has a role and function as a solution to environmental issues including flood disasters through teaching to students as the next generation of the nation with the content of hydrosphere material. This research aims to find out the influence of hydrosphere material knowledge on the attitude of high school students in conducting water conservation in Brebes Regency. The method used in the study was a survey using questionnaires consisting of questions to measure knowledge levels and a set of statements to measure students' attitudes towards water conservation. The survey was conducted on 270 respondents from nine high schools in Brebes Regency. Data analysis in this study used simple linear regression to determine the influence between knowledge and attitudes. The results showed a positive influence of hydrosphere material knowledge on the attitude of students in water conservation in Brebes Regency by 3.4%, with the value of t-count > t-table, that is, t-count by 3091 > t-table 1969. Based on the results of the data analysis, it can be concluded that 96.6% of students' attitudes towards water conservation are influenced by other factors that are not studied by the authors.

Keywords Knowledge of hydrosphere materials · Attitude of students · Water conservation

1 Introduction

Educational institutions are one of the determinants of the quality of the next generation of Indonesia. Education in school institutions takes place during the learning process and interaction of various components of the school beyond classroom learning. This learning process is very dependent on the curriculum that applies in

Ristiani (✉) · R. Dede · S. Iwan
Universitas Pendidikan Indonesia, Bandung City, West Java, Indonesia
e-mail: ristiani19@upi.edu; dede_rohmat@upi.edu; iwansetiawan@upi.edu

© The Author(s), under exclusive license to Springer Nature
Switzerland AG 2023
W. Boonpook et al. (eds.), *Applied Geography and Geoinformatics for Sustainable Development*, Springer Geography,
https://doi.org/10.1007/978-3-031-16217-6_2

Indonesia. Geography is included as one of the subjects in the education curriculum in Indonesia.

One of the sub-subjects of geography in the tenth grade high school, namely, about hydrosphere dynamics and their impact on life, in the aspect of knowledge, teaches students to understand the importance of hydrosphere and water conservation and in the aspect of attitude demanding students to be able to be critical and concerned about water issues that occur in the surrounding environment and be positive about water conservation efforts made. Meanwhile, in the aspect of behaviour/skills, students are taught to apply water conservation efforts in the surrounding environment. Thus, in this case, the learning of geography plays an important role.

However, based on the observations of the author, these learning goals were not fully achieved, by looking at the fact that there are still many natural disasters caused by humans, for example, flood events in several sub-districts in Brebes Regency, including Tanjung, Bulakamba, Wanasari, Brebes, Bantarkawung, Bumiayu and Sirampog. Based on these issues, there needs to be water conservation. Water conservation activities require integrated efforts from various parties, including educational institutions in which there are human resources in the form of students.

This research will focus on high school students in Brebes Regency. The behaviour of water conservation students can be identified through the attitude of students in doing water conservation, namely, the tendency of students to care or not care about the urgency of water conservation. The attitude of students in relation to water conservation is important to evaluate to know the success of geography learning in schools. In addition, adolescents according to the *World of Health Organization* (WHO) (in Diananda [7, p. 117]) are a population in the age range of 10–19 years old, which is the age of transition from child to adult accompanied by the process of self-discovery, so they are relatively emotionally unstable. Therefore, education is very essential in this phase, so that in the future students can become a society that has responsibility for the environment.

This research is based on Ajzen theory [3] (in Seni and Ratnadi [22, p. 4046]) which is about *planned behaviour theory* and explains that attitudes can determine behaviour. So, the behaviour of students in later life in the society can be estimated through attitudes in the present in the learning or education process.

Randolph and Troy's [18, p. 441] research states that concern for global climate change issues and the way water resources are exploited influences water conservation attitudes and actions. The same opinion was expressed by Clark and Finley [6, p. 623] that people who have a broad perspective on global climate issues where it has local impacts will be easier to do water conservation. Roseth [21] (in Gilbertson et al. [10, p. 57]) argues that water conservation is easier if there is knowledge on the availability of water reserves being used. However, the studies did not examine the influence of climate change knowledge on water conservation behaviour. Climate is included in the scope of hydrosphere material in the tenth grade high school. Therefore, this study aims to examine how knowledge of hydrosphere material affects the attitude of high school students in conducting water conservation. It is also based on the inequality of conditions between the role of geography and the real conditions of the research environment, where research areas are often flooded.

Based on the exposure of the background of the study, the authors are interested in analysing the influence of hydrosphere material knowledge (X) on the attitude of high school students in doing water conservation in Brebes Regency (Y).

2 Methods

2.1 Research Design

The research methods used in this study were surveys with quantitative approaches. The determination of this method aimed to find data on the influence of hydrosphere material knowledge on the attitude of high school students in conducting water conservation in Brebes Regency.

2.2 Sampling Techniques

The sample in this study was taken randomly in stratified clusters using the stratified cluster random sampling technique, namely, sampling which was carried out if the population consists of different strata and clusters [25].

The survey analysis unit in the study was a number of students as representatives of high schools in Brebes Regency impacted by the floods. Some districts impacted by flooding in Brebes Regency include Tanjung, Bulakamba, Wanasari, Brebes, Bantarkawung, Bumiayu and Sirampog. Respondents are numbered 270 people from nine (nine) schools in the sub-district impacted by the floods. Each school took 30 students majoring in IPS (Social Science) who had obtained hydrosphere material learning as representatives in answering research questionnaires.

2.3 Data Collection Techniques

The data collection technique in the study used multiple-choice tests and questionnaires. The multiple choice questions in this study were compiled to measure the knowledge of high school students in Brebes Regency regarding the mastery of hydrosphere material. The knowledge of students was measured from the ability of students in remembering (C1), understanding (C2), analysing (C4) and evaluating (C5) something related to hydrosphere material.

Meanwhile, the questionnaire used in this study was a statement (closed questionnaire) using the Likert scale. Questionnaires with Likert scale were used to measure the attitude of high school students in conducting water conservation in the flood area of Brebes Regency with weight range values 1–5 and with a choice of statements 'Strongly Disagree', 'Disagree', 'Neutral', 'Agree' and 'Strongly Agree'.

2.4 Instrument Development

The development of instruments in this study was carried out using several indicators outlined from variables to measure knowledge of hydrosphere materials and the attitude of students towards water conservation in Brebes Regency. These indicators were shown in Tables 1 and 2.

Students were asked to fill out the questions of the hydrosphere material multiple choice test as many as 40 questions referred to from the operational words C1 (remember), C2 (understand), C4 (analyse) and C5 (evaluate) with scoring criteria 0 and 1. In addition, students were also asked to fill out a questionnaire regarding attitudes towards water conservation as many as 36 statements with scoring criteria 1–5.

Table 1 Variable indicators of hydrosphere material knowledge

Variable	Variable dimensions	Indicators
Knowledge of hydrosphere material	Remember (C1)	a. Recognize b. Remember
	Understand (C2)	a. Interpret b. Exemplify c. Classify d. Conclude e. Explain
	Analyse (C4)	a. Distinguish b. Attribute/connect/associate
	Evaluate (C5)	a. Check b. Criticize

Source: Anderson et al. [4]

Table 2 Attitude indicators of high school students in conducting water conservation

Variable	Indicators
Student attitude	a. General attitude towards water conservation b. Past behaviour c. Moral obligation to conserve water resources d. Behavioural intentions in conserving water resources e. Nuanced analysis of water resources f. Perception of the right to water g. Past experience

Source: Reddy et al. [19]

2.5 Data Analysis Techniques

The data analysis technique in this study used simple linear regression analysis to find the influence of hydrosphere material knowledge on the attitude of high school students in conducting water conservation in the flood areas of Brebes Regency. The equations for simple linear regression were as follows [30]:

$$Y = a + bX \tag{1}$$

where Y is the dependent variable, a is the regression constant and bX is the derivative/increase value of the independent variable.

The data obtained from respondents was still ordinal data, so the data needed to be converted into interval data through *the successive interval method* by Statcal. The requirement of simple linear regression analysis was that the instrument should be valid and reliable. In addition, the data also needed to be tested first with a classical assumption test. The classical assumption test aimed to determine the ability of the regression coefficient as the best unbiased estimator (*Best Linear Unbiased Estimator/BLUE*) [23, p. 109]. The classical assumption test as a requirement to perform a simple linear regression analysis was *the Kolmogorov-Smirnov* normality test, *the linearity deviation from linearity* test and *Spearman's Rho* heteroscedasticity test. Moreover, the model feasibility test was carried out to ensure that the regression model was feasible or not to be used through model reliability test (F test), correlation coefficient test (r) and coefficient of determination test (r^2). Meanwhile, the hypothesis test in this study used the t-test by comparing the t-count and t-table values.

3 Results and Discussion

3.1 Validity and Reliability

Analysis of the influence of hydrosphere material knowledge on the attitude of high school students in water conservation began with a validity and reliability test of multiple-choice tests and questionnaires with IBM SPSS 24. The validity of the multiple choice questions was seen from the *Pearson correlation*/value of r-count. Here were the validity test results of multiple choice tests and questionnaires.

Based on Fig. 1, it could be known that there were two questions out of 40 multiple choice questions that were declared invalid. Next, the author used the drop method and tests the validity of the second stage. Here were the results of the validity test about the second stage (Fig. 2).

From the second stage of multiple choice validity test results, it could be known that 38 multiple choice questions remain valid and worthy to be included in the reliability test. Here were the reliability tests on multiple choice.

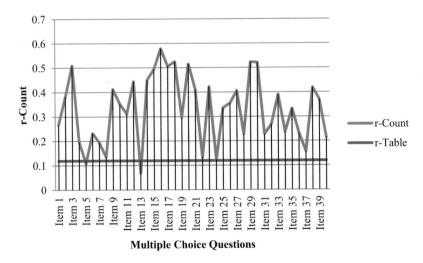

Fig. 1 The first stage of multiple choice validity test results

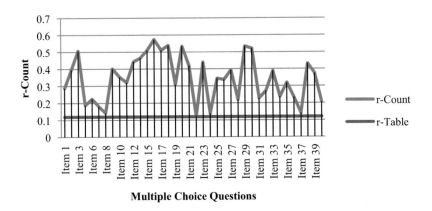

Fig. 2 The second stage of multiple choice validity test results

The basis of reliable decision-making was seen from the value of Cronbach's alpha, provided that if the value of Cronbach's alpha was greater than 0.70, the instrument was reliable. Table 3 shows Cronbach's alpha value of 0.801, hence the question of reliable multiple choice. Then the level of reliability was categorized according to the provisions in Table 4.

Based on Cronbach's alpha value of 0.801, the reliability level of multiple choice was very high. Furthermore, questionnaires of students' attitudes towards water conservation in the study were also tested for validity. The results of the question-naire validity test could be seen in Fig. 3.

From Fig. 3, it could be known that all statements in the questionnaire of students' attitudes regarding water conservation were declared valid. Meanwhile,

Table 3 Reliability test results on multiple choice

Reliability statistics	
Cronbach's alpha	No. of items
.801	38

Table 4 Reliability criteria

Reliability value range	Category
$0.800 < r_1 \leq 1000$	Very high
$0.600 < r_1 \leq 0.799$	High
$0.400 < r_1 \leq 0.599$	Adequate
$0.200 < r_1 \leq 0.399$	Low
$0.00 \leq r_1 \leq 0.199$	Very low

Source: Arikunto [5]

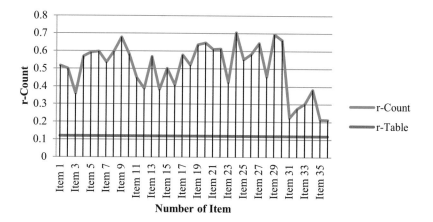

Fig. 3 Results of student attitude questionnaire validity test

Table 5 Results of the reliability test of the attitude questionnaire of students

Reliability statistics	
Cronbach's alpha	No. of items
.900	36

the results of the reliability test of the student attitude questionnaire could be seen in Table 5.

When compared, Cronbach's Alpha value of 0.900 was greater than 0.70 meaning it's a reliable questionnaire to be used as a measuring instrument. The level of reliability of the questionnaire according to the provisions in Table 4 on reliability criteria, the questionnaire of the attitude of students, was classified as having a very high level of reliability.

3.2   Classical Assumption Test

The classical assumption test was performed as a condition for simple linear regression analysis. Classical assumption tests in simple linear regression include normality tests, linearity tests and heteroscedasticity tests.

3.2.1   Normality Test

The normality test in this study used *the Kolmogorov-Smirnov* test. The basis of decision-making in this normality test was that if the significance value >0.05, then the residual value is normal distribution. Whereas if the significance value <0.05, then the residual value does not distribute normally. The results of the normality test in this study were shown in Table 6.

Based on the results of the normality test known significance values of 0.200 > 0.05, it can be concluded that residual values are normal distribution.

3.2.2   Linearity Test

The linearity test in this study used a decision-making basis based on sig. Values of deviation from linearity with the provision if sig. Value of deviation from linearity >0.05, and then there was a linear relationship between the hydrosphere material knowledge variable and the student's attitude variable. If the sig. Value of deviation from linearity <0.05, then there was no linear relationship between the hydrosphere material knowledge variable and the attitude variables of high school students in water conservation. Here was a table of linearity test results (Table 7).

Table 6 *Kolmogorov-Smirnov* normality test results

One-sample Kolmogorov-Smirnov test		
		Unstandardized residual
N		270
Normal parameters[a,b]	Mean	.0000000
	Std. deviation	15.83182806
Most extreme differences	Absolute	.037
	Positive	.025
	Negative	−.037
Test statistic		.037
Asymp. Sig. (two-tailed)		.200[c,d]

[a]Test distribution is normal
[b]Calculated from data
[c]Lilliefors significance correction
[d]This is a lower bound of the true significance

Table 7 Linearity test results based on sig. value of deviation from linearity

ANOVA table

			Sum of squares	df	Mean square	F	Sig.
The Attitudes of High School Students in Conducting Water Conservation (Y) * Hydrosphere Material Knowledge (X)	Between groups	(Combined)	61732.784	237	260.476	1.030	.484
		Linearity	2404.086	1	2404.086	9.503	.004
		Deviation from linearity	59328.697	236	251.393	.994	.536
	Within groups		8095.286	32	252.978		
	Total		69828.070	269			

Table 8 Results of *Spearman's Rho* heteroscedasticity test

Correlations

			Hydrosphere Material Knowledge (X)	Unstandardized residual
Spearman's Rho	Hydrosphere Material Knowledge (X)	Correlation coefficient	1.000	−.035
		Sig. (two-tailed)	.	.571
		N	270	270
	Unstandardized residual	Correlation coefficient	−.035	1.000
		Sig. (two-tailed)	.571	.
		N	270	270

Based on the results of the linearity test known sig. value of deviation from linearity 0.536 > 0.05, it can be concluded that there is a linear relationship between knowledge of hydrosphere material and the attitude of high school students in water conservation.

3.2.3 Heteroscedasticity Test

The heteroscedasticity test in this study used Spearman's Rho test with the provision if sig. value (two-tailed) > 0.05 then the conclusion was that there were no symptoms of heteroscedasticity. Conversely, if the sig. value (two-tailed) < 0.05, then the conclusion was that there occurred symptoms of heteroscedasticity. Here were the results of the heteroscedasticity test.

Table 8 above showed sig. value (two-tailed) of knowledge of hydrosphere material (X) was 0.571 > 0.05. So in conclusion there are no symptoms of heteroscedasticity because the significance value is greater than 0.05 (statistical confidence level of 95% or 0.05).

The results of classical assumption tests showed normal distributed data, there was a linear relationship between knowledge of hydrosphere material and the attitude of high school students in water conservation, and there are no symptoms of heteroscedasticity. That is, data is eligible for analysis through simple linear regression analysis.

3.3   Simple Linear Regression Analysis

This research is based on the theory that knowledge has a significant effect on attitudes. A simple linear regression analysis was performed to test the influence between hydrosphere material knowledge variables (X) and the attitude of high school students in doing water conservation (Y). The results of a simple linear regression analysis with IBM SPSS 24 could be seen in Table 9.

Based on Table 9 above, it was known that constant value (a) of 95.242, while the knowledge value of hydrosphere material (b/regression coefficient) was 0.292, so that the regression equation can be written:

$$Y = a + bX \tag{2}$$

$$Y = 95.242 + 0.292X$$

where X is the knowledge of hydrosphere material and Y is the attitude of high school students in doing water conservation.

The equation means that a constant of 95.242 was a fixed value of the student's attitude variable (Y) at the time the value of the hydrosphere material knowledge variable (X) was 0 or not increased. Meanwhile, the regression coefficient value (b) of 0.292 (positive) indicates a direct or positive influence, which means if the knowledge of hydrosphere material (X) is increased by one unit it will increase the attitude of high school students in doing water conservation (Y) by 0.292 units.

After the value of the simple linear regression equation was known, then the model feasibility test was carried out to ensure that the regression model was feasible or not to be used through model reliability test (F test), correlation coefficient test (r) and coefficient of determination test (r^2). Meanwhile, the hypothesis test in

Table 9 Results of simple linear regression analysis

Coefficients[a]						
	Unstandardized coefficients		Standardized coefficients			
model	B	Std. error	Beta		t	Sig.
1 (Constant)	95.242	6.912			13.780	.000
Hydrosphere Material Knowledge (X)	.292	.094	.186		3.091	.002

[a]Dependent variable: The Attitudes of High School Students in Conducting Water Conservation (Y)

this study used the t-test by comparing the t-count and t-table values. The results of the F test analysis with SPSS could be seen in Table 10.

The ANOVA table shows that the F number is 9.556 with a significance level of 0.002 (Sig. $F <= 0.05$). This means that the regression model generated from the independent variable knowledge of the hydrosphere material and the dependent variable of the attitude of high school students in carrying out water conservation is declared good and very feasible (goodness of fit).

Significance determination of the relationship, the direction of the relationship and the level of strength of the relationship between the variables of knowledge of the hydrosphere material and the variables of students' attitudes towards water conservation are seen from the correlation coefficient. The correlation coefficient test in this study used Pearson correlation analysis with SPSS. Furthermore, to be able to give an interpretation of the strength or weakness of the influence, the guidelines for interpreting the correlation coefficient in Table 11 are used.

Meanwhile, the results of the Pearson correlation analysis could be seen in Table 12.

Based on the results of the SPSS analysis, it can be seen that the value of Sig. (two-tailed) is worth $0.002 < 0.05$, so it can be concluded that the knowledge (X) and attitude (Y) variables are correlated. Pearson correlation value is 0.186. The correlation coefficient is positive (+), so it can be concluded that the relationship between the knowledge variable and the attitude variable is unidirectional. That is, if there is an increase in the knowledge variable, the attitude variable will also increase and vice versa.

Meanwhile, if seen from Table 11, the Pearson correlation value is included in the 0.00–0.199 interval with a very low level of relationship. So it can be concluded that the level of strength of the relationship between knowledge of the hydrosphere

Table 10 *F* test results based on the ANOVA table

ANOVA[a]

Model		Sum of squares	df	Mean square	F	Sig.
1	Regression	2404.086	1	2404.086	9.556	.002[b]
	Residual	67423.984	268	251.582		
	Total	69828.070	269			

[a]Dependent variable: The Attitudes of High School Students in Conducting Water Conservation (*Y*)
[b]Predictors: (constant), Hydrosphere Material Knowledge (*X*)

Table 11 Interpretation of correlation coefficient

Coefficient interval	Relationship level
0,00–0,199	Very low
0,20–0,399	Low
0,40–0,599	Medium (strong enough)
0,60–0,799	Strong
0,80–1000	Very strong

Source: Sugiyono [24]

Table 12 Pearson correlation analysis results

Correlations

		Hydrosphere Material Knowledge (X)	The Attitudes of High School Students in Conducting Water Conservation (Y)
Hydrosphere Material Knowledge (X)	Pearson correlation	1	.186**
	Sig. (two-tailed)		.002
	N	270	270
The Attitudes of High School Students in Conducting Water Conservation (Y)	Pearson correlation	.186**	1
	Sig. (two-tailed)	.002	
	N	270	270

** Correlation is significant at the 0.01 level (two-tailed)

Table 13 Determination coefficient test results

Model summary[b]

Model	R	R square	Adjusted R square	Std. error of the estimate
1	.186[a]	0.034	0.031	15.861338

[a]Predictors: (constant), Hydrosphere Material Knowledge (X)
[b]Dependent variable: The Attitudes of High School Students in Conducting Water Conservation (Y)

material and the attitude of high school students in carrying out water conservation is very low.

Furthermore, the coefficient of determination (r^2) or the determining coefficient shows the magnitude of the influence or contribution of the knowledge variable (X) to the rise/fall of the attitude variable (Y). The coefficient of determination (r^2) is obtained from the correlation coefficient squared and expressed in percent. Analysis of the coefficient of determination (r^2) in this study uses SPSS. The coefficient of determination can be seen in Table 13.

Based on the table, it can be seen that the value of R square (r^2) is 0.034. The value of the coefficient of determination is obtained with the following results:

$$\text{Coefficient of Determination} = (r)^2 \times 100\%$$
$$= 0.034 \times 100\%$$
$$= 3.4\%$$

Based on these calculations, it is known that the coefficient of determination is 3.4%, which means that the variable knowledge of the hydrosphere material contributes 3.4% to the attitude variable of high school students in carrying out water conservation. While the remaining 96.6% is caused by other factors not examined by the author.

Referring to the values of Sig. F, r and r^2, it can be concluded that this regression model is feasible and good, so that the parameter values generated by this regression model are correct, accurate and scientifically reliable to be used in analysing the effect of hydrospheric material knowledge on attitudes of high school students in carrying out water conservation in Brebes Regency.

Furthermore, hypothesis testing is carried out to ensure that the regression coefficient (increase or decrease value) is significant or not in the sense that the hydrospheric material knowledge variable (X) affects the attitude variable of high school students in carrying out water conservation (Y). Table 9 shows that the t-count of the knowledge variable is 3.091, while the t-table value is determined with a significance level of 5% (0.05) and the degree of freedom df $= (n-k)$ is 268.

The basis for decision-making in hypothesis testing is simple linear regression analysis, among others, by looking at the comparison between t-count and t-table. If the value of t-count > t-table, it means that the knowledge variable (X) has an effect on the attitude variable (Y). Meanwhile, if the value of t-count < t-table, it means that the knowledge variable (X) has no effect on the attitude variable (Y). So with a significance level of 5% (0.05) and df $= n-2 = 268$, the t-table value is 1.969. Because t-count $= 3.091 >$ t-table $= 1.969$, it can be concluded that the knowledge of hydrosphere material (X) affects the attitude variable of high school students in water conservation (Y).

Hypothesis testing is also done by comparing the significance value of the t-count knowledge variable in Table 9 with a probability value of 0.05 based on the provisions: if the knowledge significance value is <alpha 0.05, it means that the knowledge variable (X) has a significant effect on the attitude variable (Y). Meanwhile, if the knowledge significance value > alpha 0.05, it means that the knowledge variable (X) has no significant effect on the attitude variable (Y). The significance level of the knowledge variable is 0.002; thus there is a significant influence between the variables of knowledge of the hydrosphere material (X) on the attitude of high school students in carrying out water conservation (Y).

The simple linear regression model has three elements of theory testing, namely, the number b, the sign of the relationship and the level of significance. This study resulted in a regression model with a value of $b \neq 0$, an equal sign with theory (positive) and sig. < 0.05. This means that knowledge of the hydrosphere material affects the attitude of high school students in carrying out water conservation in the same direction as the theory, which is positive. Based on the three elements of testing this theory, it can be concluded that the results of this study are the influence of knowledge of the hydrosphere material on the attitudes of high school students in carrying out water conservation in Brebes Regency, supporting and strengthening the theory of the influence of knowledge on attitudes.

The results of the analysis of the influence of knowledge of the hydrosphere material on the attitudes of high school students in carrying out water conservation in Brebes Regency are in accordance with the results of the research of Kusuma and Untarini [15, p. 1579] that knowledge has a significant effect on attitudes. The same results were also found in the research of Rohmatun and Dewi ([20, p. 33]. The better the knowledge and understanding of students, the knowledge will be stored in the mind and will significantly increase the positive attitude of students towards the environment [17, p. 63].

This research is also in line with the findings of Wolters [29], Estrada [9], Vassileva [27], Gregory and Leo [11], Syme et al. [26] and Aitken et al. [2] in Enshassi et al. [8, p. 166] that knowledge in this case, especially knowledge about controlling and limiting water use, can increase changes in attitudes and behaviour towards water conservation. As stated by the research conducted by Hines et al. [12] and Hungerford and Volk [14] in Wang and Wang [28, p. 7] that a number of studies have confirmed that cognitive factors, such as knowledge of ecological or environmental concepts and environmental sensitivity or awareness influence environmental attitudes and belief values, thereby increasing environmental behaviour. Another supporting statement, namely, research by McGuinness et al. [16, p. 383], found the fact that there was a significant influence between knowledge, attitudes and behaviour on the environment. The three variables are directly related to each other; in other words, if knowledge increases, attitudes also increase, as well as behaviour will show an increase.

Other suitable research findings are according to Hsu and Feng [13, p. 181] that environmental knowledge helps individuals to develop positive attitudes towards the environment. Knowledge directly affects the level of behavioural control and attitudes towards pro-environmental behaviour [1, p. 1356].

The results showed that the attitude of students in carrying out water conservation in Brebes Regency was very good, while the mastery of hydrospheric material knowledge in most schools was still lacking. However, as stated by Aertsens et al. [1, p. 1368] that a more positive attitude can lead to higher levels of interest and thus higher levels of knowledge. The results of these studies strengthen the researcher's hypothesis that knowledge and attitudes towards the environment, especially water conservation, influence each other.

4   Conclusion

The effect of knowledge of the hydrosphere material on the attitude of high school students in carrying out water conservation in Brebes Regency is 3.4%. The simple linear regression model obtained is $Y = 95.242 + 0.292X$. This regression model is feasible to use based on the results of the F test with a significance level of 0.002 (Sig. $F <= 0.05$). The relationship between knowledge and attitude variables is very low with a Pearson correlation value of 0.186. Meanwhile, the results of hypothesis testing based on the comparison of the t-count value with the t-table with a significance level of 5% and df = 268 obtained the t-count value = 3.091 > t-table = 1.96 9, and so it can be concluded that the variable knowledge of the hydrosphere material (X) has an effect on the attitude variable of high school students in carrying out water conservation (Y). Hypothesis testing using a t-count significance value with a probability of 0.05 also states that the significance level of the knowledge variable is 0.002 < 0.05, so it can be concluded that there is a significant influence between the hydrospheric material knowledge variable (*X*) on the attitude of high school students in carrying out water conservation (*Y*).

Conflict of Interests The authors state that there is no conflict of interest with any financial, personal, other persons or organization related to the material in this study.

References

1. Aertsens, J., Mondelaers, K., Verbeke, W., Buysse, J., Huylenbroeck, G.V.: The influence of subjective and objective knowledge on attitude, motivations and consumption of organic food. Br. Food J. **113**(11), 1353–1378 (2011)
2. Aitken, C. K., McMahon, T. A., Wearing, A. J., & Finlayson, B. L.: Residential water use: Predicting and reducing consumption 1. J. Appl. Soc. Psychol. **24**(2), 136–158 (1994)
3. Ajzen, I.: Attitudes, Personality, and Behavior-Icek Ajzen Google Books. New York, NY: Open University Press. Albrechtsen (2005)
4. Anderson, L.W., Krathwohl, D.R., Airasian, P.W., Cruikshank, K.A., Mayer, R.E., Pintrich, P.R., et al.: In: Burvikovs, A.E. (ed.) A Taxonomy for Learning, Teaching, and Assessing: A Revision of Bloom's Taxonomy of Educational Objectives. Longman (2001)
5. Arikunto, S.: Dasar-dasar Evaluasi Pendidikan, 3rd edn. PT Bumi Aksara, Jakarta (2018)
6. Clark, W.A., Finley, J.C.: Determinants of water conservation intention in Blagoevgrad, Bulgaria. Soc. Nat. Resour. **20**(7), 613–627 (2007)
7. Diananda, A.: Psikologi Remaja dan Permasalahannya. ISTIGHNA. **1**(1), 116–133 (2018)
8. Enshassi, A., Elzebdeh, S., Mohamed, S.: Drivers affecting household residents' water and related energy consumption in residential buildings. Int. J. Build. Pathol. Adapt. **35**(2), 159–175 (2017)
9. Estrada, M.: Every drop counts: UC Berkeley student water conservation behavior. Available at: nature.berkeley.edu/classes/es196/projects/2013final/EstradaM_2013. pdf (accessed 29 May 2017) (2013)
10. Gilbertson, M., Hurlimann, A., Dolnicar, S.: Does water context influence behaviour and attitudes to water conservation? Australas. J. Environ. Manag. **18**(1), 47–60 (2011)
11. Gregory, G. D., & Leo, M. D.: Repeated behavior and environmental psychology: the role of personal involvementand habit formation in explaining water consumption 1. J. Appl. Soc. Psychol. **33**(6), 1261–1296 (2003)
12. Hines, J. M., Hungerford, H. R., & Tomera, A. N.: Analysis and synthesis of research on responsible environmentalbehavior: A meta-analysis. J. Environ. Educ. **18**(2), 1–8 (1986)
13. Hsu, J.L., Feng, C.-H.: Evaluating environmental behaviour of the general public in Taiwan: implications for environmental education. Int. J. Comp. Educ. Dev. **21**(3), 179–189 (2019)
14. Hungerford, H. R., & Volk, T. L.: Changing learner behavior through environmental education. J. Environ. Educ. **21**(3), 8–21 (1990)
15. Kusuma, I.D., Untarini, N.: Pengaruh Pengetahuan Produk Terhadap Niat Beli dengan Sikap Sebagai Variabel Intervening. Jurnal Ilmu Manajemen. **2**(4), 1573–1583 (2014)
16. McGuinness, J., Jones, A.P., Cole, S.G.: Attitudinal correlates of recycling behavior. J. Appl. Psychol. **62**(4), 376–384 (1977)
17. Novarita, A., Sugandhi, D., Pasya, G.K.: Peranan Pembelajaran Geografi dalam Pembentukan Sikap Peserta Didik Terhadap Mitigasi Bencana Gempa Bumi dan Longsor di Kota Padang. Gea: Jurnal Pendidikan Geografi. **15**(1), 55–63 (2015)
18. Randolph, B., Troy, P.: Attitudes to conservation and water consumption. Environ. Sci. Pol. **11**(5), 441–455 (2008)
19. Reddy, A., Lewis, C.M., Sengupta, R.: Scale for attitude towards water conservation. SocArXiv. **1–11** (2020)

20. Rohmatun, K.I., Dewi, C.K.: Pengaruh Pengetahuan dan Religiusitas Terhadap Niat Beli pada Kosmetik Halal Melalui Sikap. Jurnal Ecodemica. **1**(1), 27–35 (2017)
21. Roseth, N.: Community views on water shortages and conservation. Water: Journal of the Australian Water Association, **33**(8), 62–66 (2006)
22. Seni, N.N., Ratnadi, N.M.: Theory of Planned Behavior untuk Memprediksi Niat Berinvestasi. E-Jurnal Ekonomi dan Bisnis Universitas Udayana. **6**(12), 4043–4068 (2017)
23. Silalahi, D., Hulu, E.: Indikator Kolektibilitas Kredit Joint Financing Menggunakan OLS & Logit. Jurnal Ilmu Keuangan dan Perbankan (JIKA). **11**(1), 107–123 (2021)
24. Sugiyono: Metode Penelitian Kuantitatif Kualitatif dan R&D. Alfabeta, Bandung (2012)
25. Sukmadinata, N.S.: Metode Penelitian Pendidikan, 10th edn. PT Remaja Rosdakarya, Bandung (2015)
26. Syme, G. J., Nancarrow, B. E., & Seligman, C.: The evaluation of information campaigns to promote voluntary household water conservation. Eval. Rev. **24**(6), 539–578 (2000)
27. Vassileva, I.: Characterization of household energy consumption in Sweden: energy savings potential and feedback approaches (Doctoral dissertation, Mälardalen University) (2012)
28. Wang, Y.-F., Wang, C.-J.: Do psychological factors affect green food and beverage behaviour? An application of the theory of planned behaviour. Br. Food J. **118**(9), 1–59 (2016)
29. Wolters, E. A.: Attitude–behavior consistency in household water consumption. Soc. Sci. J. **51**(3),455–463 (2014)
30. Yuliara, I.M.: Modul Regresi Linier Sederhana. Universitas Udayana, Bali (2016). Retrieved March 16, 2021, from https://simdos.unud.ac.id/uploads/file_pendidikan_1_dir/321812643899 0fa0771ddb555f70be42.pdf

Optimizing Multi-reservoir Systems with the Aid of Genetic Algorithm: Mahanadi Reservoir Project Complex, Chhattisgarh

Shashikant Verma (ID)**, A. D. Prasad, and Mani Kant Verma** (ID)

Abstract In a multi-reservoir system, a single reservoir's activity might have an impact on the other reservoirs within the system. Therefore, the reservoir must be managed as a single entity to ensure long-term water conservation. However, the integrated operation of a multi-reservoir system becomes more difficult when there is less rainfall in a basin. For this reason, the existing operating policy must be evaluated to ensure integrated functioning. Metaheuristic-based algorithms such as genetic algorithms (GA) are employed in this study to find the optimal solution, and optimization efficiency can be improved by avoiding local optimal solutions. It aims to develop a steady-state optimum operating policy using a genetic algorithm that can satisfy long-term demand and measure the performance along with a deterministic simulation-optimization model (S-O Model) in terms of reliability, resilience, vulnerability, and sustainability indices of the existing multi-reservoir system, namely, Ravishankar Sagar reservoir, Dudhawa, and Murrum Silli in Chhattisgarh, India. In addition, the overall performance of the existing reservoir is improved based on reliability by 35.66%, resilience by 40.54%, sustainability by 36.70%, and vulnerability reduced by 54.09%, respectively, compared to actual water release. Apart from that, the GA model release satisfactorily meets the needs of demand, and no deficit conditions have occurred during the entire study period, excluding the years 1989–1990, 2001–2002, and 2002–2003, respectively.

Keywords Operating policy · Multi-reservoir system optimization · Performance evaluation · Genetic algorithm · Simulation

S. Verma (✉) · A. D. Prasad · M. K. Verma
Civil Engineering Department, National Institute of Technology, Raipur, India

© The Author(s), under exclusive license to Springer Nature
Switzerland AG 2023
W. Boonpook et al. (eds.), *Applied Geography and Geoinformatics for
Sustainable Development*, Springer Geography,
https://doi.org/10.1007/978-3-031-16217-6_3

1   Introduction

The widespread availability of water resources through spatial and temporal distribution causes demand and supply deficits. Dams are built and developed to address the difficulty of water supply. Nevertheless, it is important, as a result of the changing way of life, population growth, changes to development priorities, conflict in reservoir operations, and the rise in demand, that the results of different reservoir operations should be measured and improved to meet the challenges of variability. Therefore, the performance of operational policies must be evaluated periodically to optimize water use. In addition to performance evaluation, reservoir operating policies must be optimized for optimal resource utilization [24]. Linear programming, dynamic programming, evolutionary algorithms, etc. are the most common optimization approaches nowadays. Different parameters, such as physical features of the reservoir, objective function, implicit restrictions, the insufficient flow of water into the reservoirs, and climate change, depend on the application of particular optimization techniques in the relevant area [23]. Therefore, no such optimization approach can universally be used for all the reservoirs [6]. Alternatives must be developed and compared with the risks associated with water resource planning and operational policies. The risk is normally calculated under two criteria: (i) the specified chance of unwanted events, (ii) the number of unwanted events during a specific time, and (iii) the expected number of unwanted occurrences over a specified time [5, 20, 25, 26]. Multiple reservoirs in a multi-reservoir system are connected in the same basin, either in series or in parallel. Dynamic, linear, and nonlinear programming were frequently used to optimize such a complex multi-reservoir system problem. If dynamic programming (DP) is used for the multi-reservoir systems, it does, however, have a big problem. There are essential approximation problems associated with discontinuous, non-differential, nonconvex, or multimodal objective functions, known as the "curse of dimensionality." Genetic algorithms have attracted much attention in natural adaptive systems because of their flexibility and efficacy for improving complex systems and appropriate optimization methods for multi-reservoir systems because they address multidimensional and multi-objective problems. Dimensions do not limit the GAs because computer memory grows linearly, not exponentially. There is no "dimensionality curse." Due to the point-by-point approach used by conventional optimization methods such as dynamic programming (DP), linear programming (LP), and nonlinear programming (NLP), multi-objective optimization is not appropriate for these approaches. As a result of this, GAs do not rely on a single solution for each iteration. This is one of the key differences between traditional and GA optimization strategies [11].

A simulation model is used to reproduce real-world behavior, which means it emulates reality. The most basic form of simulation is the "what-if" condition, reflecting the system's response when confronted with the described situation in the future [15]. Simulation is preferred for complicated systems in which only aspects relevant to the study's objectives are modeled. Deterministic or stochastic simulation depends on the unpredictability of variables, and uncertainty can be reduced by using deterministic models. However, if the process is stationary, stochastic modeling may be appropriate, whereas deterministic simulation is used to assess the reservoir system's performance in response to an objective function specified using restricted linear programming.

According to the literature, some studies have been done in the same study area to develop a steady-state operating policy for the multi-reservoir system. Anusha et al. [3] examined the reliability, resilience, and vulnerability of a multi-reservoir system using a deterministic simulation model that integrates constrained linear programming. Stochastic dynamic programming was utilized to develop a monthly steady-state optimum operating policy for the Ravishankar Sagar reservoir [4]. According to Ananda Babu et al. [2], particle swarm optimization (PSO) was used to develop the optimal operating policy for the Ravishankar Sagar reservoir, and they found that deficits were reduced significantly when employing the PSO algorithm. In addition, based on the previous study some relevant research gaps can be incorporated into the present study.

This study makes use of a reservoir system operating model, which has a longer time horizon and more decision variables and constraints. This algorithm's effectiveness is evaluated by comparing it to the results of an existing operational policy that represents an optimal solution. A meta-heuristic approach will be used in this work to find an optimal reservoir operating policy for a multi-reservoir system and compare it to the existing operating policy. The following is the outline for the paper. An introduction to the genetic algorithm is in order. In the following section, we briefly discuss the formulation of the objective function for the multi-reservoir optimization problem and how it operates. The next part deals with the representation of the study area and data used, the methodology, parameter selection for the genetic algorithm, model development, and its findings, as well as some general observations and inferences to be made.

2   Study Area and Data Used

The Ravishankar Sagar reservoir is a reservoir commonly referred to as Gangrel Dam in Chhattisgarh, India. It has been built on the opposite bank of the Mahanadi river. It is located roughly 17 km from Dhamtari and 90 km from Raipur in the Dhamtari district. It is the largest structure for storing water in Chhattisgarh, and it is used for both industrial and household irrigation [19, 27]. Table 1 shows the silent features of the multi-reservoir system, including storage reservoir constraints, and Fig. 1 represents the study area map of the multi-reservoir system.

Table 1 Silent features of Ravishankar Sagar reservoir

Parameters	Quantities
Type of dam	Earth fills embankment
Impoundment	Mahanadi river
Total capacity	910.50 MCM
Active capacity	766.890 MCM
Dead storage	144 MCM
Surface area	95 Sq. km
Normal elevation	333 m
Height	30.50 m
Length	1830 m

Source: https://en.wikipedia.org/wiki/Gangrel_Dam

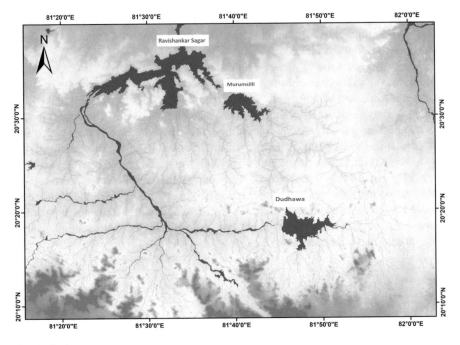

Fig. 1 Study area map

2.1 Data Used

Detailed information on the reservoir characteristics such as rainfall data, inflow, and outflow of the Ravishankar Sagar reservoir is available for 31 years, from 1989–1990 to 2019–2020. Irrigation water is supplied from the Mahanadi Main Canal (MMC) and Mahanadi Feeder Canal (MFC). Whereas water is supplied to meet the monthly household water demands of Raipur, Durg, and Dhamtari, respectively, for domestic purposes, the industrial demand is supplied to Bhilai Steel Plant, Bhilai. The annual total water supplied for industrial and household demand is 238.0 Mm^3.

3 Model Formulation

3.1 Fitness Function

The optimization model and fitness function used in the present study reduce the total monthly square deviation from specified release and demand and the square deviation from the mass balance equation between June and July to as little as possible. The objective function is as follows:

$$Z = Minimize \sum_{i=1}^{3} \sum_{t=1}^{12} \left(Rt - Dt \right)^2 \tag{1}$$

where, $i = 1, 2, 3$ and $n = 1, 2, 3 ....., 12$, Rt = release from the reservoir at time t, and Dt = demand required for the reservoir at time t.

3.2 Constraints

Non-negative constraints:

$$R_{1,t} + R_{2,t} + R_{3,t} \geq 0 \tag{2}$$

where $R_{1,t}$ = release from reservoir 1 at time t and similarly $R_{2,t}$ and $R_{3,t}$ are the releases from reservoirs 2 and 3 at time t, respectively.

1. Mass balance equation:

$$S_t = S_{t+1} + I_t - R_t - E_t - O_t \tag{3}$$

where S_t = initial storage at the beginning of the month, S_{t+1} = final storage at the end of the month, I_t = monthly inflow during the specific period, E_t = monthly evaporation loss, and O_t = monthly overflow loss.

2. Supply constraints:

$$\left. \begin{cases} 0 \leq S_{1,t} \leq D_{1,t} \\ 0 \leq S_{2,t} \leq D_{2,t} \\ 0 \leq S_{3,t} \leq D_{3,t} \end{cases} \right\} \tag{4}$$

where $S_{1,t}$ = storage for reservoir 1 at time t, and similarly $S_{2,t}$ and $S_{3,t}$ are the storage for reservoirs 2 and 3 at time t, respectively, and $D_{1,t}$ = demand from reservoir 1 at time t, and similarly $D_{2,t}$ and $D_{3,t}$ are the demand from reservoirs 2 and 3 at time t, respectively.

3. Release constraints:

$$S_{1,t} + S_{2,t} + S_{3,t} \leq R_{1,t} + R_{2,t} + R_{3,t} \tag{5}$$

Continuity equation:

$$\begin{aligned}
(\text{Storage})_{R,\max\_(\text{live},t)} &\geq (\text{Storage})_{t-1} + I_{3,t} - S_{1,t} - S_{2,t} + S_{3,t} + R_{1,t} - R_{2,t} \\
&- (\text{Overflow losses})_t - (\text{Evaporation})_{R,t} \geq (\text{Storage})_R
\end{aligned} \tag{6}$$

4. Storage constraints

The storage reservoir must not exceed the reservoir's capacity in any specific month and must not go below the dead capacity, as mentioned:

$$S\min \leq St \leq S\max \tag{7}$$

where $Smin$ = dead storage of the reservoir in Mm^3, S_t = storage of the reservoir in Mm^3, and $Smax$ = maximum storage of the reservoir in Mm^3.

4   Methodology

4.1   Genetic Algorithm

A genetic algorithm (GA) model was designed for a multi-reservoir system in Chhattisgarh, India. It is a random search algorithm based on natural selection and genetic principles. GA modeling is gaining in popularity due to its ability to perform a robostic random search and produce near-global optimal values. GA was formed in the mid-1970s due to robostic random search and comparatively near-optimal values [9]. Many researchers, recently, summarized genetic algorithm simulation [14, 21, 28]. Most research has focused on applying GA to groundwater and water distribution network problems. However, little work has been done on applying GA to reservoir operational problems and watershed management [1, 8, 11, 12, 14, 16, 22]. The objective was to maximize the benefits of power generation, and storage and release constraints apply to agricultural water supplies and drinking water supplies [14]. GA may be used to develop appropriate operating policies for multi-reservoir systems, as shown in this paper. A genetic algorithm was employed to optimize a multi-reservoir system in Indonesia (Brantas Basin) [22]. Kim and Heo [11] used multi-objective GAs to optimize the multi-reservoir system, and a population curve defining optimal solutions was derived. There is limited use of GA in multi-objective, multi-reservoir system optimization and to determine the appropriate operating policy for a multi-purpose reservoir and compare its performance to SOP [1].

A GA model was developed to guide operational policies for a multi-reservoir system. The main purpose of this study is to develop a steady-state optimum operating policy that can meet long-term demand and compare the genetic algorithm model performance with deterministic simulation-optimization models (S-O Models) in terms of reliability, resilience, sustainability, and vulnerability of existing multi-reservoir systems. Reservoir capacity, evaporation, release, and mass balance constraints govern the water released for agriculture and other uses. The fitness of each string is coded into strings, and the decision variables have specified upper and lower bounds. These strings are reproduced, crossed, and mutated. This section presents a broad methodology for the model formulation for optimization of a multi-reservoir system. The following steps constitute the most frequent optimization procedure, as indicated in Fig. 2:

1. Problem formulation.
2. Categorization of the decision variable.
3. The objective function and constraint limits are used to set limits on the values of certain decision variables.
4. The decision variable's values must be finalized.
5. Steps 3 and 4 are repeated until the stop criterion has been satisfied. Step 3 is where the algorithms differ the most from the previous steps. In this step, the search mechanism of the algorithm is different.

4.2   Reservoir Performance Evaluations

4.2.1   Reliability

The reliability is based on the possibility that a reservoir can meet the target demand at any given time interval during the simulation period. The system's reliability may be represented as a frequency or probability α that satisfies the system state [10]:

$$\alpha = \text{Prob}\big[X_t \epsilon S\big] \tag{8}$$

where X_t = system output state at time t and S = group of all satisfactory outputs. Reliability is sometimes thought of as the exact opposite of risk.

4.2.2   Resilience

Resilience indicates how quickly a system recovers from failure. The likelihood of a successful year after a failed year is called resilience [10].

$$\text{Res} = \left\{ \frac{\frac{1}{M} \sum_{j=1}^{M} d_j}{\frac{1}{M}} \right\} \tag{9}$$

where d_j = duration of the failure event and M = number of failure events.

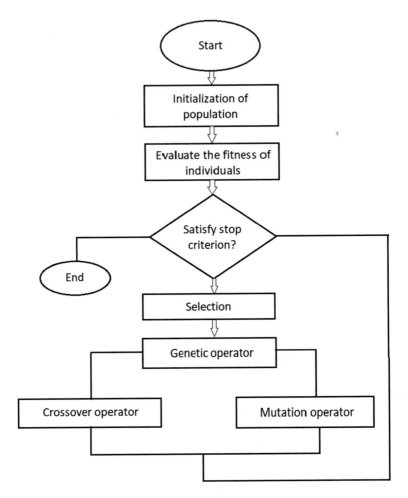

Fig. 2 Procedure for GA model

4.2.3 Vulnerability

The likelihood of damage caused by a failed event is measured as vulnerability. Vulnerability is defined as the mean of the deficiency events over a while. It is given as:

$$\text{Vul} = \frac{1}{M}\sum_{j=1}^{M} v_j \tag{10}$$

where v_j = the failure event deficit volume, M = number of failure events, and Vul = vulnerability index.

4.2.4 Sustainability Index

Sustainability indexes have been proposed by [13] and include reliability, resilience, and vulnerability ratios. It is expressed as:

$$k = \text{Re}_t \times \text{Res}\left(1 - \frac{\text{Vul}}{D}\right) 0 \leq k \leq 1 \tag{11}$$

where k = sustainability index, D = target demand, Re_t = reliability index, Res = resilience index, and Vul = vulnerability index.

4.3 Model Performance Criteria

To check the accuracy of the employed algorithm, the statistical evaluation indexes of the coefficient of determination (R^2), root mean squared error (RMSE), mean absolute percentage error (MAPE), and normalized mean squared error were used as per the equation from (12) to (15) [29]:

$$R^2 = \left[\frac{\sum\left(\text{Re}_{\text{opt}(i)} - \overline{\text{Re}_{\text{opt}}}\right)\left(\text{Re}_t - \overline{\text{Re}}\right)}{\sqrt{\sum\left(\text{Re}_{\text{opt}(i)} - \overline{\text{Re}_{\text{opt}}}\right)^2 \sum\left(\text{Re}_t - \overline{\text{Re}}\right)^2}}\right]^2 \tag{12}$$

$$\text{RMSE} = \sqrt{\frac{1}{n}\sum_{t=1}^{n}\left(\text{Re}_{\text{opt}(i)} - \text{Re}_t\right)^2} \tag{13}$$

$$\text{MAPE} = \frac{100}{n}\sum_{i=1}^{n}\left|\frac{\text{Re}_{\text{opt}(i)} - \text{Re}_t}{\text{Re}_{\text{opt}(i)}}\right| \tag{14}$$

$$\text{NMSE} = \frac{\sum_{i=1}^{n}\left(\text{Re}_t - \text{Re}_{\text{opt}(i)}\right)^2}{N\sum_{i=1}^{n}\left(\text{Re}_t\right)^2} \tag{15}$$

In the above equation, Re_t is release in time t for the optimized algorithm, $\overline{\text{Re}}$ is the mean of the release, $\text{Re}_{\text{opt}(i)}$ is optimum release at time t, and similarly $\overline{\text{Re}_{\text{opt}}}$ is the mean of optimum release and n is the number of total periods.

The lower the RMSE and the higher the R^2 value, the more accurate the algorithm is and the better it is compared to other algorithms. An algorithm's MAPE and NMSE are a measurement of the difference between its release and its current state, and hence its lower value indicates a better result.

5   Results and Discussions

5.1   Sensitivity Analysis

Since GA relies on fine-tuning its primary parameters, these studies were carried out using crossover probability (P_c), mutation probability (P_m), the number of search population, and the number of iterations, whereas the parameters were adjusted within the following range $P_c = [0.5–0.95]$, $P_m = [0.01–0.09$ and $0.1–1.0]$ [7, 17, 18]. In the present study the selected parameters are $P_c = 0.9$, $P_m = 0.7$, $n = 30$, and $I_{max} = 1000$. The crossover and mutation probability GA model can get the optimum value of the objective function at $P_c = 0.9$ and $P_m = 0.7$. The algorithm execution time increases and decreases the objective function value after each alteration of the parameter's value. Similarly, the GA model can get the optimal value of an objective function after 1000 iterations. Each additional iteration increases the algorithm execution time and decreases the objective function value. Therefore, the GA algorithm goes through a total of 1000 iterations. For the multi-reservoir system, these values can be shown in Fig. 3a–c, respectively.

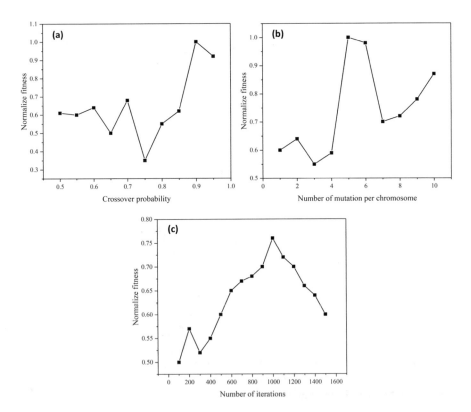

Fig. 3 (**a**), (**b**), and (**c**). Sensitivity analysis for parameter adjustment. (Source: own study)

5.2   Optimization Operation Policy

The genetic algorithm results have been used to derive operation policy results for a study area from 1989–1990 to 2019–2020. Monthly inflows of Ravishankar Sagar, Dudhawa, and Murrum Silli reservoirs are provided as inputs, and monthly deficits and monthly releases of the reservoirs are assessed as outputs. The algorithm was solved in MATLAB, and the actual deficit results achieved by running the algorithm from 1989–1990 to 2019–2020 are presented in graphical form with the time t on the X-axis (years) and releases (Mm³) on the Y-axis. Figure 4 illustrates the existing operating policy of a multi-reservoir system and the release policy produced by the genetic algorithm. As seen in Fig. 5, the GA model's key advantage is its ability to find solutions that are close to being globally optimal. Only in 1989–1990, 2001–2002, and 2002–2003, respectively, have there been water shortages when the reservoir system operation model is compared to the water release under existing policy. Due to rigorous release and demand planning, the total squared deviations decreased. Concerning both the release and demand of the GA model, there were

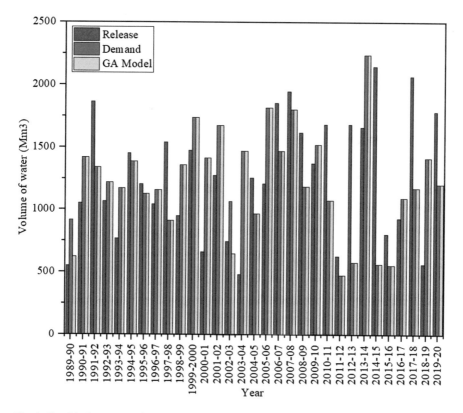

Fig. 4 Graphical representation of release made by genetic algorithm operation policy concerning existing release and demand

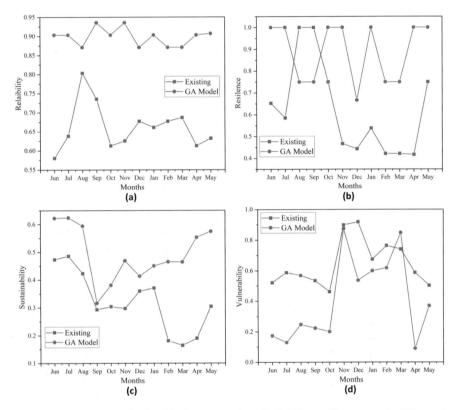

Fig. 5 (**a**), (**b**), (**c**), and (**d**). Graphical representation of reliability, resilience, sustainability, and vulnerability measures for genetic algorithm operation policy release concerning existing releases

many constraints. Therefore, the GA model has been successful in delivering sufficient monthly releases for multi-reservoir systems compared to the existing reservoir operating policy.

5.3 Performance Measures

Based on historical data, the simulation and optimization model were run for 31 years (372 months) from June 1989–1990 to May 2019–2020. Simulation model performance is assessed as per Sect. 4.3 by computing several performance evaluation criteria, as shown in Table 2. Figure 5a–d illustrate graphical representations of reliability, resilience, sustainability, and vulnerability indices for genetic algorithm operation policy releases concerning existing releases. The GA model outperforms existing operational policy in terms of performance, increasing average percentage changes to 12.79%, 3.98%, and 15.48% for reliability, resilience, and sustainability,

Table 2 Model evaluation criteria for the multi-reservoir system

Case study	Algorithm	Evaluation criteria			
Multi-reservoir system	Genetic algorithm	R^2	RMSE	MAPE	NMSE
		0.9454	1.007	0.0594	0.0545

respectively, while vulnerability reduces up to 10.28%. Four statistical evaluation criteria, as previously stated in Sect. 4.3, were utilized to compare the performance of the GA model in multi-reservoir systems. According to Table 2 results, the GA model had the maximum R^2 in the multi-reservoir system (0.9454), yet it had the least error parameters (RMSE = 1.007, MAPE 0.0594, and NMSE = 0.0545) in the system.

6 Conclusions

A genetic algorithm optimization model has been developed and deployed to derive the optimal operating policy for the multi-reservoir system in Chhattisgarh, India, namely, Dudhawa, Murrum Silli, and Ravishankar Sagar, respectively. Release and target demand were designed to minimize the sum of squared deviations. The GA model was subject to several constraints, including those related to release and demand. The primary benefit of the GA model is its robust search, which frequently yields near-global optimal solutions. Compared to the existing operations, the GA model has yielded no deficit conditions apart from 1989 to 1990, 2001 to 2002, and 2002 to 2003, respectively. Compared to the existing reservoir operation policy, the GA model successfully provided appropriate monthly releases for the Ravishankar Sagar reservoir. The overall performance of the Ravishankar Sagar reservoir is improved based on an average, reliability of 35.66%, resilience of 40.54%, sustainability of 36.70%, and vulnerability reduced to 54.09%, respectively.

Acknowledgments The authors would like to extend their admiration and respect to NIT, Raipur, Chhattisgarh.

References

1. Ahmed, J.A., Sarma, A.K.: Genetic algorithm for optimal operating policy of a multi-purpose reservoir. Water Resour. Manag. **19**(2), 145–161 (2005). https://doi.org/10.1007/s11269-005-2704-7
2. Ananda Babu, K., Shrivastava, R.K., Dikshit, M.: Optimal operation policies for Ravishankar Sagar reservoir – a case study. J. Indian Water Resour. Soc. **35**(1) (2015)
3. Anusha, N., Verma, M.K., Bajpai, S.: Performance measures for Ravi Shankar Sagar Reservoir using simulation-optimization models. Water Util. J. **15**, 67–79 (2017)
4. Anusha, N., Verma, M., Bajpai, S.: Developing of monthly steady state optimum operating policies for Ravishankar Sagar Reservoir using stochastic Dynamic Programming. In: Indian Water Week-2016 (2016)

5. Azharuddin, M., Verma, S., Verma, M.K., Prasad, A.D.: A synoptic-scale assessment of flood events and ENSO – streamflow variability in Sheonath River Basin, India. In: Rao, C.M., Patra, K.C., Jhajharia, D., Kumari, S. (eds.) Advanced Modelling and Innovations in Water Resources Engineering Lecture Notes in Civil Engineering, vol. 176, pp. 93–104. Springer, Singapore (2022). https://doi.org/10.1007/978-981-16-4629-4_8

6. Chauhan, S., Shrivastava, R.K.: Performance evaluation of reference evapotranspiration estimation using climate-based methods and artificial neural networks. Water Resour. Manag. 23(5), 825–837 (2009). https://doi.org/10.1007/s11269-008-9301-5

7. Deb, K., Agrawal, S., Pratap, A., Meyarivan, T.: A fast elitist non-dominated sorting genetic algorithm for multi-objective optimization: NSGA-II. In: International Conference on Parallel Problem Solving from Nature, pp. 849–858 (2000). https://doi.org/10.1007/3-540-45356-3_83

8. Dhiwar, B.K., Verma, S., Prasad, A.D.: Identification of flood vulnerable area for Kharun River Basin by GIS techniques. In: Rao, C.M., Patra, K.C., Jhajharia, D., Kumari, S. (eds.) Advanced Modelling and Innovations in Water Resources Engineering Lecture Notes in Civil Engineering, vol. 176, pp. 385–408. Springer, Singapore (2022). https://doi.org/10.1007/978-981-16-4629-4_27

9. Forrest, S.: Genetic algorithms: principles of natural selection applied to computation. Science. 261(5123), 872–878 (1993). https://doi.org/10.1126/science.8346439

10. Hashimoto, T., Stedinger, J.R., Loucks, D.P.: Reliability, resiliency, and vulnerability criteria for water resource system performance evaluation. Water Resour. Res. 18(1), 14–20 (1982). https://doi.org/10.1029/WR018i001p00014

11. Kim, T., Heo, J.H.: Multireservoir system optimization using multi-objective genetic algorithms. In: Critical Transitions in Water and Environmental Resources Management, pp. 1–10 (2004). https://doi.org/10.1061/40737(2004)244

12. King, J.P., Fahmy, H.S., Wentzel, M.W.: A genetic algorithm approach for river management. In: Dasgupta, D., Michalewicz, Z. (eds.) Evolutionary Algorithms in Engineering Applications. Springer, Berlin/Heidelberg (1997). https://doi.org/10.1007/978-3-662-03423-1_7

13. Loucks, D.P.: Quantifying trends in system sustainability. Hydrol. Sci. J. 42(4), 513–530 (1997). https://doi.org/10.1080/02626669709492051

14. Oliveira, R., Loucks, D.P.: Operating rules for multireservoir systems. Water Resour. Res. 33(4), 839–852 (1997). https://doi.org/10.1029/96WR03745

15. Patel, P.L., Sharma, P.J.: Hydroclimatic variability across Tapi Basin, India: issues and implications. In: Climate Change-Sensitive Water Resources Management, pp. 46–64 (2020). https://doi.org/10.1201/9780429289873-5

16. Pradhan, D., Sahu, R.T., Verma, M.K.: Flood inundation mapping using GIS and Hydraulic model (HEC-RAS): a case study of the Burhi Gandak River, Bihar, India. In: Kumar, R., Ahn, C.W., Sharma, T.K., Verma, O.P., Agarwal, A. (eds.) Soft Computing: Theories and Applications Lecture Notes in Networks and Systems, vol. 425, pp. 135–145. Springer, Singapore (2022). https://doi.org/10.1007/978-981-19-0707-4_14

17. Pradhan, N.S., Tripathy, K.U.: Optimization of the operating policy of the multipurpose Hirakud Reservoir by Genetic Algorithm. Am. J. Eng. Res. 2(11), 174–184 (2013)

18. Reddy, M.J., Kumar, D.N.: Optimal reservoir operation using multi-objective evolutionary algorithm. Water Resour. Manag. 20(6), 861–878 (2006). https://doi.org/10.1007/s11269-005-9011-1

19. Sahu, R.T., Verma, M K , Ahmad, I.: Segmental Variability of Precipitation in the Mahanadi River Basin During 1901–2017. PREPRINT (Version 1) available at Research Square (2021a). https://doi.org/10.21203/rs.3.rs-542786/v1

20. Sahu, R.T., Verma, M.K., Ahmad, I.: Regional frequency analysis using L-moment methodology – a review. In: Pathak, K.K., Bandara, J.M.S.J., Agrawal, R. (eds.) Recent Trends in Civil Engineering Lecture Notes in Civil Engineering, vol. 77, pp. 811–832. Springer, Singapore (2021b). https://doi.org/10.1007/978-981-15-5195-6_60

21. Savic, D.A., Walters, G.A.: Genetic algorithms for least-cost design of water distribution networks. J. Water Resour. Plan. Manag. **123**(2), 67–77 (1997). https://doi.org/10.1061/(ASCE)0733-9496(1997)123:2(67)
22. Sharif, M., Wardlaw, R.: Multireservoir systems optimization using genetic algorithms: case study. J. Comput. Civ. Eng. **14**(4), 255–263 (2000). https://doi.org/10.1061/(ASCE)0887-3801(2000)14:4(255)
23. Sudhagar, P.E., Babu, A.A., Vasudevan, R., Jeyaraj, P.: Vibration analysis of a tapered laminated thick composite plate with ply drop-offs. Arch. Appl. Mech. **85**(7), 969–990 (2015). https://doi.org/10.1007/s00419-015-1004-9
24. Vedula, S., Mohan, S., Shrestha, V.S.: Improved operating policies for multipurpose use: a case study of Bhadra Reservoir. Sadhana. **9**(3), 157–176 (1986). https://doi.org/10.1007/BF02811963
25. Verma, S., Prasad, A.D., Verma, M.K.: Trend analysis and rainfall variability of monthly rainfall in Sheonath River Basin, Chhattisgarh. In: Recent Trends in Civil Engineering, pp. 777–790. Springer, Singapore (2021). https://doi.org/10.1007/978-981-15-5195-6_58
26. Verma, S., Prasad, A.D., Verma, M.K.: Trends of rainfall and temperature over Chhattisgarh during 1901–2010. In: Rao, C.M., Patra, K.C., Jhajharia, D., Kumari, S. (eds.) Advanced Modelling and Innovations in Water Resources Engineering Lecture Notes in Civil Engineering, vol. 176, pp. 3–19. Springer, Singapore (2022a). https://doi.org/10.1007/978-981-16-4629-4_1
27. Verma, S., Sahu, R.T., Prasad, A.D., Verma, M.K.: Development of an optimal operating policy of multi-reservoir systems in Mahanadi Reservoir Project Complex, Chhattisgarh. International conference on applications of intelligent computing in engineering and science. J. Phys. Conf. Ser. **2273**, 012020 (2022b). https://doi.org/10.1088/1742-6596/2273/1/012020
28. Wardlaw, R., Sharif, M.: Evaluation of genetic algorithms for optimal reservoir system operation. J. Water Resour. Plan. Manag. **125**(1), 25–33 (1999). https://doi.org/10.1061/(ASCE)0733-9496(1999)125:1(25)
29. Willmott, C.J.: On the validation of models. Phys. Geogr. **2**(2), 184–194 (1981). https://doi.org/10.1080/02723646.1981.10642213

Monitoring of Morphological Change in Lam Phachi River Using Geo-informatics System

Thanat Saprathet, Chudech Losiri, Asamaporn Sitthi,
and Jeerapong Laonamsai 🔘

Abstract Nowadays, economic and industrial development along the Lam Phachi River in Ratchaburi and Kanchanaburi provinces is accelerating and increasing water demand. Therefore, weirs were built to preserve the water demand, and the sand extraction businesses were also carried out to support economic expansion. These activities resulted in river morphological change relative to the past. Additionally, climate change has resulted in a rapid change in river discharge and riverbank erosion each season. To monitor the morphology of the river, normalized difference water index (NDWI), soil adjusted vegetation index (SAVI), and a maximum likelihood classification were used to analyze the pattern using remotely sensed images obtained from the Thaichote satellite between 2013 and 2015 and the Sentinel-2A satellite between 2017 and 2019. The results indicated that the Lam Phachi River follows a dendritic drainage pattern, with similar characteristics of riverbank erosion in all directions. The rivers obtained from the NDWI extraction had an accuracy of 94.11% and an analytical precision of 95.15%. The SAVI results indicated an accuracy of 93.17% and a precision of 94.33%. The maximum likelihood classification results showed an accuracy of 96.67% and an analytical precision of 97.47%. The stream extraction method based on the maximum likelihood classification is the most accurate. It was found that the erosion was a lot in the middle of the river, and most of the deposition areas are found at the end of the river. However, the NDWI and SAVI extractions are less accurate in this study area due to water hyacinth and a slight difference in water content between the riverbank and the water bodies.

T. Saprathet · C. Losiri (✉) · A. Sitthi
Department of Geography, Faculty of Social Sciences, Srinakharinwirot University,
Bangkok, Thailand
e-mail: chudech@g.swu.ac.th; asamaporn@g.swu.ac.th

J. Laonamsai
Department of Civil Engineering, Faculty of Engineering, Naresuan University,
Phitsanulok, Thailand

Keywords Remote sensing · SAVI · NDWI · Maximum likelihood classification ·
Riverbank erosion · River morphology

1   Introduction

Water is a vital resource on the planet that all organisms require to survive. It is a
significant world component, accounting for up to three-fourths of its surface.
Although water is abundant, population growth and economic expansion have
increased the water demand. Due to environmental and climatic changes, natural
disasters such as floods, droughts, landslides, and soil erosion are becoming more
frequent and severe [1]. Additionally, industrial expansion contributes to the esca-
lating water pollution problem in the modern era [2].

Thailand has encountered water resource issues on various dimensions, including water quality, drought, flooding, and riverbank erosion [3]. Riverbank erosion is
primarily caused by river curvature, discharge, near-surface wind speed, and waves
[4]. Floods affect geographical change, such as riverbank erosion, because the force
of the river flow exceeds the riverbank's resistance capacity. In addition, human-
caused bank erosion occurs because of transportation and the sand mining industry
[5]. The Lam Phachi River originates in the Tenasserim Mountains in western
Thailand. Because of its curvature and geological setting, the Lam Phachi River has
been eroded [6]. As a result, the water consumed by humans is muddy and sedimentary, posing health risks and affecting agricultural products. Sand vacuuming also
results in the collapse of riverbanks and alters the flow of the river [7].

The issues mentioned above in the Lam Phachi River are critical. They must be
resolved immediately, and they may have a long-term impact on natural resources
and humans, including the livelihoods and utilities of people living along the river.
The remote sensing method has been widely used to monitor the eroded and deposited areas of the river [8, 9]. Thus, the current study used remote sensing to investigate river morphology and riverbank erosion and deposition in the Lam Phachi
River. Thaichote and Sentinel-2A satellite images were used in this study. Three
methods including the soil adjusted vegetation index (SAVI), the normalized difference water index (NDWI), and the maximum likelihood image classification were
used to extract the river morphological change. This study will contribute valuable
information about river morphology and bank erosion to the relevant government
organization to develop sustainable river management policies.

2   Study Area

The Lam Pachi River basin is located in Ratchaburi and Kanchanaburi provinces
and covers an area of 2550 km². The Lam Pachi River flows through Ratchaburi
Province from its source in the Tenasserim Hills. In Kanchanaburi Province, the

river merges with the Mae Klong River. The majority of the Lam Phachi basin is located in Ratchaburi, while a small portion is located in Kanchanaburi. The main channel measures 130 km in length. The river runoff flows from the south to the north. Stream flow is seasonal in nature, with two distinct seasons: wet and dry. October has the highest discharge, averaging 56 m³/s, while January through March has the lowest, averaging less than 3 m³/s [10]. Additionally, the most significant environmental problems are related to river flow during the two previous seasons: flooding during the wet season and drought during the dry season (Fig. 1).

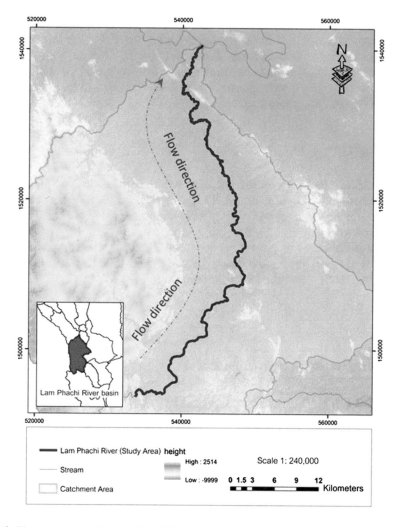

Fig. 1 The study area in the Lam Phachi River basin

3   Data and Method

3.1   Satellite Images

3.1.1   Thaichote Satellite

Thaichote, as known as THEOS, was wholly owned and operated by the Thai Ministry of Science and Technology's Space Agency (GISTDA). It provided Thailand with globally georeferenced image products and image processing capabilities for use in cartography, land use, agricultural monitoring, forestry management, coastal zone monitoring, and flood risk management [11]. It was intended to be a small satellite weighing 750 kg. It was launched into low earth orbit in 2008, reaching an altitude of 822 km with a 98-degree inclination [12]. It is equipped with optical imagery capable of detecting objects in the visible to near-infrared range. The image resolution is 15 m perpendicular to the earth's surface, with a width of 90 km. The Thaichote satellite images are smaller than the study area. Three images were mosaiced using the Geographic Information System (ArcMap) program to ensure the data's completeness and continuity.

3.1.2   Sentinel-2A

Sentinel-2A is a European optical imaging satellite launched in 2015 at the height of 786 km as part of the Copernicus Programme of the European Space Agency [13]. The satellite is equipped with a wide-swath, high-resolution multispectral imager with a resolution of 10–60 m and 13 spectral bands [14]. It conducts global-terrestrial observations to support services such as forest monitoring, detection of land cover changes, and natural disaster management. Table 1 details the satellite images used for Thaichote and Sentinel-2A.

Table 1 Details of satellite images used in this study

Satellite	Date	Month	Year	Path/row	Source
Thaichote	4	February	2013	150/35	GISTDA
	28	February	2015		
Sentinel-2A	20	January	2017	R104	USGS
	26	November	2019		

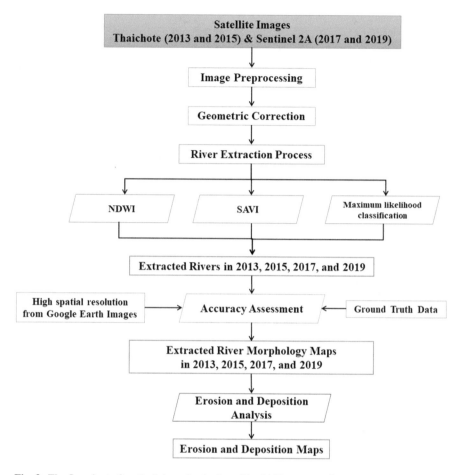

Fig. 2 The flowchart of methodology for the Lam Phachi River extraction

3.2 River Extraction Method

To extract the Lam Phachi River from the satellite images, this study designed the overall methodology as shown in Fig. 2. After downloading the satellite image from the online data archives, all images were passed into data preprocessing processes such as atmospheric correction, geometric correction, and image mosaicking. Then, all images were also resampled to 15-meter spatial resolution.

To analyze satellite images, three different methods were used to extract stream-lines: the soil adjusted vegetation index (SAVI), the normalized difference water index (NDWI), and the maximum likelihood image classification method. The streamlines from all three methods were combined to create a riverbank line that is complete and accurate, with more efficient and precise data for the analysis.

3.2.1   Normalized Difference Water Index (NDWI)

The normalized difference water index (NDWI) is used to highlight open water features in satellite images by utilizing the NIR (near-infrared) and GREEN (visible green) spectral bands [15]. The index is given as:

$$NDWI = (GREEN - NIR) / (GREEN + NIR) \tag{1}$$

NDWI is an appropriate vegetation index for distinguishing land from water. This is because the surface of the water has a high absorption of electromagnetic radiation and low radiation reflectance. There is, however, an error in estimating the construction site as water. As a result, it is crucial to determine the ratio of NIR and GREEN wavelengths. The NDWI value is between −1 and 1, with 1 indicating the presence of water bodies or extremely high humidity and − 1 indicating a dry area or no moisture [16]. The NDWI was calculated for this study using the raster calculator toolbox in the ArcMap program. The NDWI value between 0.19 and 0.58 indicates the presence of a water body, as illustrated in Fig. 3a.

3.2.2   Soil Adjusted Vegetation Index (SAVI)

NDWI products derived empirically are inherently unstable, varying with soil color, soil moisture, and saturation effects caused by dense vegetation [17]. As a result, the SAVI was created to account for the differential extinction of red and near-infrared light within the vegetation canopy. The index is a transformation technique that reduces the influence of spectral vegetation indices that use red (RED) and near-infrared (NIR) wavelengths on soil brightness [18]. The SAVI index is given as:

$$SAVI = ((NIR - -RED) / (NIR + RED + L)) \times (1 + L) \tag{2}$$

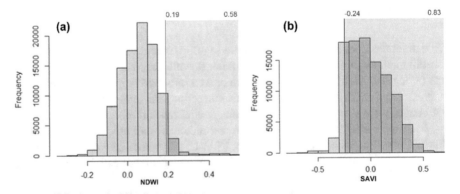

Fig. 3 Histogram of (**a**) NDWI value and (**b**) SAVI value

where L is a factor for adjusting the canopy background. The L value of 0.5 was discovered in reflectance space to minimize soil brightness variations and eliminate the need for additional calibration for different soils [19]. It was found that the transformation virtually eliminates soil-induced variations in vegetation indices. The study extracts satellite imagery data with SAVI using the red and near-infrared bands via a raster calculator in the ArcMap program. The SAVI values ranged from −0.24 to 0.83 (Fig. 3b).

3.2.3 Image Classification Method

The training area was defined using supervised classification and shapefiles to visualize the river line. The maximum likelihood classifier was used to extract the water body, which classifies the data by considering each data type's mean and covariance matrix [20]. For this study, satellite image data, including visible and near-infrared wavelengths, were extracted via classification of image data using ArcMap's functional classification [21]. It begins by selecting a representative sample of the training sites, which are the water areas. It is assumed that each data type has a normal distribution and then determines which data type has the highest probability of containing each image point.

3.3   Data Validation

A confusion matrix summarizes prediction results on image classification [22]. Correct and incorrect predictions are outlined and broken down by class using count values. Next, this study compared the observed classification set to the predicted classification set. Four distinct outcomes are possible in any given column [23, 24], as illustrated in Table 2. Firstly, the classifier correctly identified the water sample. This is referred to as a true positive (TP). Secondly, the classifier incorrectly classifies the water sample as land or vegetation, resulting in a false-negative (FN) result. Thirdly, the classifier misclassifies the land sample as water. This condition is referred to as a false-positive result (FP). Fourthly, the classifier correctly identifies the land sample as a true negative result (TN).

By overlaying the riverbanks from the Google Earth base map on the river line data extracted from satellite imagery, this study validates the river line data extracted from satellite imagery. To begin, actual points were chosen to compare streamline

Table 2 Confusion matrix

Predicted/actual	Yes	No
Yes	TP	FP
No	FN	TN

classifications for each method applied to the area. Finally, the result was incorporated into the confusion matrix, as shown in Table 2. The following accuracy and precision values were calculated:

$$\text{Accuracy} = (TP + TN) / (TP + TN + FP + FN) \tag{3}$$

$$\text{Precision} = TP / (TP + FP) \tag{4}$$

4 Results and Discussion

4.1 Streamline Extraction

Figure 4 shows the streamlines extracted from Thaichote and Sentinel-2A satellite images in 2013, 2015, 2017, and 2019 using the NDWI, SAVI, and image classification. The result indicates that for each year of interest, the streamlines extracted using these three methods are not significantly different (Fig. 4).

To validate the results, the 100 sample points of the Lam Phachi riverbank from the high-resolution images of the Google Earth [25] and a ground survey were utilized and summarized in a confusion matrix, including checks for accuracy and precision (Table 3). According to the confusion matrix, the streamline obtained from the NDWI extraction was 95% accurate and 97% precise. The river obtained through the SAVI extraction was 91% accurate and 93% precise. The streamline extracted via image classification produced the most accurate result. According to Table 3, the accuracy is 97%, and the precision is 98%. It was followed by the NDWI and SAVI extraction results.

4.2 River Morphology

The Lam Phachi River flows from south to north (Fig. 1). The morphology of the upstream river is dominated by rock, has a steep slope, and results in a relatively narrow V-shaped valley [26]. Additionally, weirs were constructed, giving the river the appearance of a meander downstream of the weir. The area between the midstream and downstream is flat. As a result, the riverbank was eroded, and the convex bank was deposited. The depositional bank was formed by accumulating coarse sand sediment and gravel. This mechanism alternates the ridge with a swale, referred to as a sand shoal, developing the Lam Phachi River's waterways. The streams form a network defined as the dendritic drainage pattern [27]. Dendritic drainage patterns are prevalent in rocky terrain. The riverbank erosion characteristics are similar in all directions. As a result, an overlay technique was used each year to obtain representative river lines (Fig. 5).

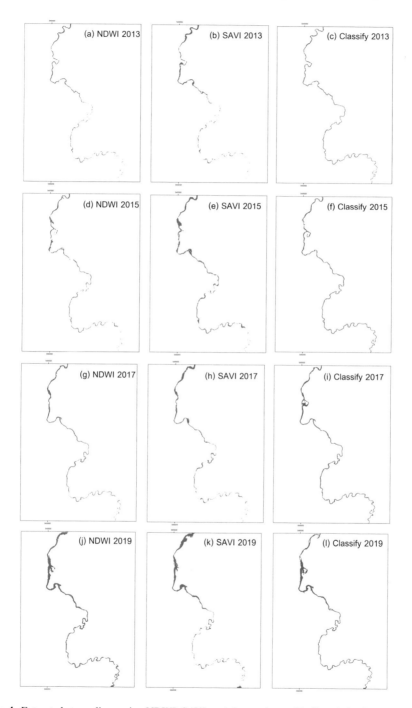

Fig. 4 Extracted streamlines using NDWI, SAVI, and the maximum likelihood classification

Table 3 Confusion matrix of the results from NDWI, SAVI, and image classification

Method	Actual Prediction	Water	Others	Accuracy (%)	Precision (%)
NDWI	Water	81	3	95	97
	Others	2	14		
SAVI	Water	90	3	91	93
	Others	6	1		
Maximum likelihood classification	Water	92	2	97	98
	Others	1	5		

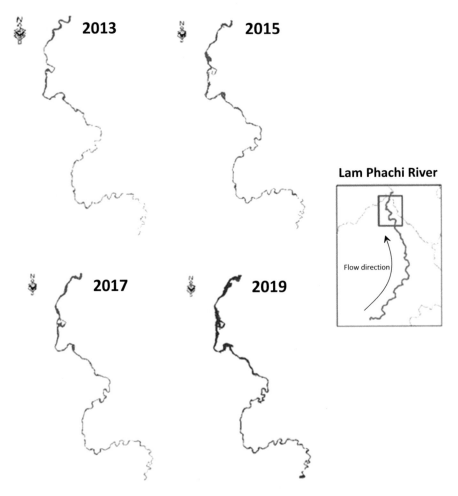

Fig. 5 An overlayed streamline between NDWI, SAVI, and maximum likelihood classification methods during (**a**) 2013, (**b**) 2015, (**c**) 2017, (**d**) 2019

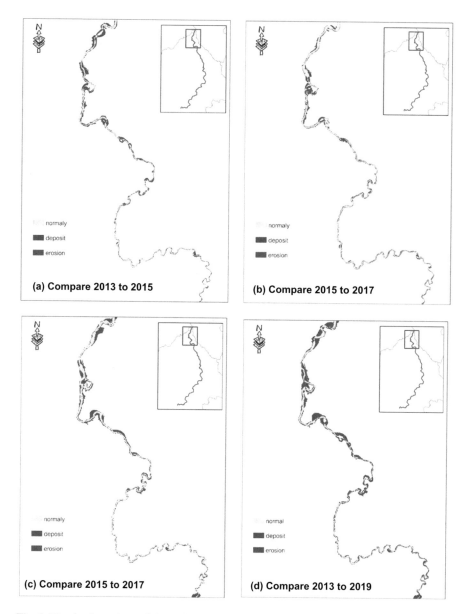

Fig. 6 Riverbank erosion and deposition along the Lam Phachi river between (**a**) 2013 and 2015, (**b**) 2015 and 2017, (**c**) 2017 and 2019, and (**d**) 2013 and 2019

4.3   Riverbank Erosion and Deposition

This study used the overlay technique to analyze changes in river morphology. Streamlines were overlaid and combined from three different methods (the maximum likelihood classification, NDWI, and SAVI) to represent the better quality of the river morphology in 2013, 2015, 2017, and 2019. Each period was examined for changes in river morphology. Figure 6a depicts riverbank erosion and deposition from 2013 to 2015. By overlapping the data for eroded (red) and deposited (green) areas along the Lam Phachi River, it is possible to calculate that the eroded area equals 0.89 km^2 and the deposition area equals 1.11 km^2. Considering the changes in river morphology between 2015 and 2017, the result indicates that the erosional area is 0.91 km^2 and there are 0.85 km^2 of depositional area (Fig. 6b).

Additionally, the streamlines between 2017 and 2019 were overlayed (Fig. 6c). The erosional area is estimated to be 1.52 km^2. The depositional area is approximately 0.34 km^2 in size. Compared to the previous period of 2013–2017, the erosion is more pronounced due to dam operation. Additionally, erosion and deposition along the Lam Phachi River were considered for the entire period of 2013–2019 (Fig. 6d). Erosion had a greater impact than deposition. Erosion totaled 1.87 km^2. The area affected by the deposition is 0.67 km^2. These findings confirm that landlords have been losing their fields at an alarming rate [7].

5   Conclusion

This study uses the NDWI, the SAVI, and the maximum likelihood classification to reflect progress in the river morphology in Lam Phachi River between 2013 and 2019 using Sentinel 2-A and Thaichote satellites. The extracted streamline results indicated that the maximum likelihood classification method yielded the most accurate with a 97% accuracy and 98% precision. Additionally, the streamline from that method contained the complete characteristics of the river. On the other hand, the SAVI and NDWI produced less accurate results because some sections of the river line were missing, particularly those with narrow channels. Therefore, the Lam Phachi River's stream patterns are described as dendritic, with similar erosion characteristics along the riverbank in all directions.

The erosion and deposition of the Lam Phachi River were analyzed from the extracted streamlines. The erosion and deposition could be found along the river. Between 2013 and 2019, an area of 1.88 km^2 was eroded. The most eroded zones were in the middle and downstream of the river, where a sand mining industry and dam operation existed. As a result, the river curved back and forth across the landscape as it flowed over gently sloping terrain. It made the river meandering. In comparison, between 2013 and 2019, the landfill area was 0.67 km^2. It was because the river discharge was low and the watercourse has changed direction, the most of the deposited land was in the middle of the Lam Phachi River.

References

1. Ahuja, S.: Lessons learned from water disasters of the world, in Separation Science and Technology, pp. 417–427. Elsevier (2019)
2. Pashtoon, R., Zaki, Z., Haqbin, N.: Empirical study on international tourism and economic growth of Thailand: An ARDL-ECM bounds testing approach. J. Enterp. Dev. **4**(1), 28–48 (2022)
3. Laonamsai, J., Ichiyanagi, K., Patsinghasanee, S.: Isotopic temporal and spatial variations of tropical rivers in Thailand reflect monsoon precipitation signals. Hydrol. Process. **35**(3), e14068 (2021)
4. Grove, R., Croke, J., Thompson, C.: Quantifying different riverbank erosion processes during an extreme flood event. Earth Surf. Process. Landf. **38**(12), 1393–1406 (2013)
5. Lusiagustin, V., Kusratmoko, E.: Impact of sand mining activities on the environmental condition of the Komering river, South Sumatera. 2017. AIP Publishing LLC (2017)
6. ThaiPBS, Sand mining activity in the Lam Phachi River, https://news.thaipbs.or.th/content/282181. Last accessed 29 May 2019.
7. MGROnline, Ratchaburi residents complained that sand mining business cause riverbank erosion, https://mgronline.com/local/detail/9610000097297. Last accessed 29 May 2019.
8. Aher, S.P., Bairagi, S.I., Deshmukh, P.P., Gaikwad, R.D.: River change detection and bank erosion identification using topographical and remote sensing data. Int J Appl Inf Syst. **2**(3), 1–7 (2012)
9. Kummu, M., Lu, X., Rasphone, A., Sarkkula, J., Koponen, J.: Riverbank changes along the Mekong River: Remote sensing detection in the Vientiane–Nong Khai area. Quat. Int. **186**(1), 100–112 (2008)
10. Areerachakul, S., Junsawang, P.: Rainfall-Runoff relationship for streamflow discharge forecasting by ANN modelling. 2014. IEEE (2014)
11. Tanarat, S., Popattanachai, P., Kasetkasem, T.: THAICHOTE level 1A production using SIPRO procedure. In: 2016 13th International Conference on Electrical Engineering/Electronics, Computer, Telecommunications and Information Technology (ECTI-CON) 2016. IEEE (2016)
12. Vongsantivanich, W., R. Sachasiri, P. Navakitkanok, J. Plaidoung, D. Niammuad, P. Popattanachai. The Evolution of GISTDA Satellite Control Center. 2014
13. Gascon, F., C. Bouzinac, O. Thépaut, M. Jung, B. Francesconi, J. Louis, V. Lonjou, B. Lafrance, S. Massera, A. Gaudel-Vacaresse, Copernicus Sentinel-2A calibration and products validation status. Remote Sens. (Basel) **9**(6), p. 584 (2017).
14. Li, J., Roy, D.P.: A global analysis of Sentinel-2A, Sentinel-2B and Landsat-8 data revisit intervals and implications for terrestrial monitoring. Remote Sens. (Basel). **9**(9), 902 (2017)
15. Gao, B.-C.: NDWI—A normalized difference water index for remote sensing of vegetation liquid water from space. Remote Sens. Environ. **58**(3), 257–266 (1996)
16. McFeeters, S.K.: The use of the Normalized Difference Water Index (NDWI) in the delineation of open water features. Int. J. Remote Sens. **17**(7), 1425–1432 (1996)
17. da Silva, V.S., Salami, G., da Silva, M.I.O., Silva, E.A., Monteiro Junior, J.J., Alba, E.: Methodological evaluation of vegetation indexes in land use and land cover (LULC) classification. Geol. Ecol. Landscape. **4**(2), 159–169 (2020)
18. Huete, A.R.: A soil-adjusted vegetation index (SAVI). Remote Sens. Environ. **25**(3), 295–309 (1988)
19. Qi, J., Chehbouni, A., Huete, A.R., Kerr, Y.H., Sorooshian, S.: A modified soil adjusted vegetation index. Remote Sens. Environ. **48**(2), 119–126 (1994)
20. Strahler, A.H.: The use of prior probabilities in maximum likelihood classification of remotely sensed data. Remote Sens. Environ. **10**(2), 135–163 (1980)
21. Gevana, D., Camacho, L., Carandang, A., Camacho, S., Im, S.: Land use characterization and change detection of a small mangrove area in Banacon Island, Bohol, Philippines using a maximum likelihood classification method. Forest Sci. Technol. **11**(4), 197–205 (2015)

22. Lewis, H., Brown, M.: A generalized confusion matrix for assessing area estimates from remotely sensed data. Int. J. Remote Sens. **22**(16), 3223–3235 (2001)
23. Visa, S., Ramsay, B., Ralescu, A.L., Van Der Knaap, E.: Confusion matrix-based feature selection. MAICS. **710**, 120–127 (2011)
24. Payne, C., Panda, S., Prakash, A.: Remote sensing of river erosion on the Colville River, North Slope Alaska. Remote Sens. (Basel). **10**(3), 397 (2018)
25. Yue, H., Li, Y., Qian, J., Lin, Y.: A new axxuracy evaluation method for water. body extraction. Int. J. Remote Sens. **41**(19), 1–32 (2020)
26. Kinnura, M., M. Higo, K. Lorsirirat, and S. Kumlungkeng, Identifying significant tributaries from human impacted sedimentary system. (2002).
27. Morisawa, M.: Distribution of stream-flow direction in drainage patterns. J. Geol. **71**(4), 528–529 (1963)

Developing Scenario of Plastic Waste Leakage in the Jakarta Hydrology Environment Using Seasonal Data Conditions and Socioeconomic Aspects

Aprilia Nidia Rinasti, Kavinda Gunasekara, Ekbordin Winijkul, Sarawut Ninsawat, and Thammarat Koottatep

Abstract Jakarta is facing the problem of plastics in the river pathways. With the higher settlement area close to the riverbanks, Jakarta should control its high potency of plastic waste leakage into the river, where 37.76% of the waste generated comes from the residential area as the primary contributor. To comply with the control mechanism, an early warning system of plastic waste outflow should be implemented. This study aims to identify plastic leakage hotspots using spatial analysis of the river before it flows to the ocean. It was conducted into three phases: source leakage identification, hydrological characterization of streamflow, and scenario analysis. In the latter the overall stages were analyzed using geographic information systems (GIS) for plastic waste distribution and location-based topographical feature analysis. Morphometric analysis is used to define the runoff from the subbasin

A. N. Rinasti (✉)
Marine Plastics Abatement Program Department of Energy, Environment, and Climate Change, Asian Institute of Technology, Khlong Luang, Pathum Thani, Thailand

Geoinformatics Center, Asian Institute of Technology, Khlong Luang, Pathum Thani, Thailand
e-mail: aprilianidia@ait.asia

K. Gunasekara
Geoinformatics Center, Asian Institute of Technology, Khlong Luang, Pathum Thani, Thailand

E. Winijkul · T. Koottatep
Marine Plastics Abatement Program Department of Energy, Environment, and Climate Change, Asian Institute of Technology, Khlong Luang, Pathum Thani, Thailand

Environmental Engineering and Management Program Department of Energy, Environment, and Climate Change, Asian Institute of Technology, Khlong Luang, Pathum Thani, Thailand

S. Ninsawat
Remote Sensing and Geographic Information Systems ProgramDepartment of Information and Communication Technologies, Asian Institute of Technology, Khlong Luang, Pathum Thani, Thailand

W. Boonpook et al. (eds.), *Applied Geography and Geoinformatics for Sustainable Development*, Springer Geography,
https://doi.org/10.1007/978-3-031-16217-6_5

65

level as the transport of leakage. Data used, including the Greater Jakarta area's waste management summary in 2020, was transformed comprehensively into the spatial distribution for detection of hotspots. Some of the spatial features include the Digital Elevation Model, population density, LULC and rainfall rate to enable the dispersal of plastic waste leakage. Through morphometric analysis, 19% of watershed resulted from the higher runoff, emphasizing the greater leakage amount into the waterbodies. Leakage hotspots contribute 7.64% of the total unmanaged plastic waste flowing into Jakarta Bay annually. By improving the seasonal scenario, the highest leakage was estimated to be 2.8 kton in February and higher in January as a result of Jakarta special social events. The result of the estimation could be used for better land utilization in the Jakarta area to minimize the plastic waste problem.

Keywords Riverbank · Plastic leakage · Morphometric analysis

1 Introduction

Plastic consumption has more than doubled since the first part of the twenty-first century [1]. It also is correlated with an increase in the population [2] because a human is a prominent plastic consumer. With the higher disposal because of the widespread use of plastic [3], the possibility of the leakage into the environment is also higher, leading to several environmental impacts [4]. Plastic pollution also could endanger water bodies, especially the ones closest to daily human activity—the rivers.

Because most plastic is generated from land-based activity [5, 6], it also can affect the marine environment through the interconnection of the river as a pathway [7–9]. The river is the main pathway of land-based plastic waste, which delivers wastes from human activity into the ocean [9, 10] and contributes more than 80% to overall marine plastic pollution [11, 12]. In addition, more activity on land could possibly change the amount of leakage into this integrated pathway to the ocean [13]. This also is correlated with the activities established for the country's development [14].

Likewise, human activity is regarded as consumer behaviour, which also contributes to the problem's primary source [15]. Contextualizing the country, the tendency toward waste disposal into the river or other water pathways has remained high [16–18]. As mentioned, one of the tendencies leading to the higher amount of waste disposed of is related to the country's development, which is associated with the disposal term Jambeck et al. used [11]. In the context of living conditions, people who live near the water pathway (i.e., river) have a tendency of target disposal into the water in underdeveloped countries [19], which is prevalent in rural areas with improper collection systems.

According to the incidence of leakage events, plastic waste disposal to the open ocean may be as high at 88% [2], which indicates the great quantity of plastics in the ocean. It also leads to the endangering of the marine ecosystem [20–23]. The relationship between high outflow from the land into the ocean indicates that action is needed to prevent more leakage. The river, as the main water pathway that flows directly to the ocean, is regarded as the most important pathway between land-based resources and the ocean [9, 24, 25]. Therefore, rivers in this context are the preventive pathway for solving marine pollution issues.

The appearance of plastic waste in the Indonesian maritime environment and its water bodies on land has become a serious issue in recent years. Based on Indonesia Marine Debris Hotspot Synthesis Report in April 2018, plastic waste leakage in the Jakarta area could amount to 10–15% of unmanaged plastic waste in the ocean [26], traveling through the river pathway. This is followed by an issue from the Egg Report in 2019 that plastic waste pollution in the river leads to persistent organic pollutants (POPs) from egg shells from poultry products in Tropodo Village, Waru Sub-district, Sidoarjo Regency, Indonesia [27]; thus, the these are indicators that plastic enters the food chain. This also is indicated by a study in Jakarta Bay that the presence of plastic becomes higher in the rainy season because of the habit of direct disposal [28, 29].

Related to the recent conditions in Indonesia with its higher level of mismanaged waste, it has the most polluted river in the world [9]; a policy and multistakeholder approach has been undertaken by the National Plastic Action Partnerships (NPAP), which developed the scenarios for the plastic circularity [30]. To improve this, more analysis of the integration between regions needs to be addressed to advance the national strategies. This study analysed plastic waste hotspots with regard to the disposal term, modelling the plastic pathway through natural conditions and projected the trend of plastic waste leakage in context with the forecasted greater occurrence of it. A spatial domain-based model was conducted to study its topographical area in the natural system. In addition, plastic waste leakage flows from the river to the estuary, which can be estimated before it enters the ocean.

As the alarming issues correlated to the river, along with the ocean, integration of a study on land and in the ocean needs to be addressed. Balancing integration between land and ocean is possible to study the ecosystem's interaction through remote-sensing technology and geographic information systems (GIS) analysis, resulting in strategic solutions. The objective of this study is to develop a new early warning system of the plastic leakage into the ocean based on geographical conditions because it explains the integration between land and ocean. With the concept of geospatial, it has the potential to assist in the detection of plastic waste and plan advanced waste management.

2 Data and Methods

2.1 Overall Methodology

Modelling of plastic waste leakage flow and its accumulation before it enters the ocean can be attained by two major work phases: calculation of plastic waste leakage and spatial analysis; and one additional phase in the scenario analysis on the emerged occurrence (shown in Fig. 1). Therefore, the workflow and data used in this

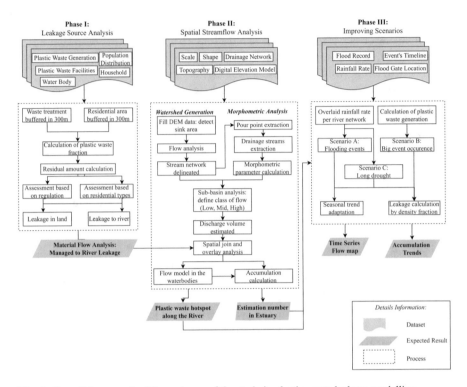

Fig. 1 Overall framework of three phases of the study in plastic waste leakage modelling

study comply with statistic and spatial data use. Both major and scenario analysis phases have various types of data processed to do a combined analysis. To obtain the modelling for a plastic waste leakage hotspot and its flow into the estuary, the workflow is divided mainly into two categories of study: material flow analysis and spatial modelling. Material flow analysis is defined as the study for plastic waste material flow from its generation point to the its outflow and discharging into the river. Spatial modelling was used as the streamflow model of the river, topographic analysis, and also the study for the material flow itself in a spatial domain. Both studies were combined for seasonal contribution analysis to conduct the estimation of a hotspot area and the monthly trends of plastic waste leakage ending up in estuaries as an accumulation.

To create the plastic waste leakage data, some spatial data distribution was considered to manage the plastic waste leakage flow—that is, household location, dumpsite location type (illegal and open dumping without proper management and inadequate for standard regulation), and location of waste treatment facilities (e.g., material recovery facility that sorted waste including plastics). The location and distribution used in this study is from 300 m away from the centroid of the river, as the study case example in Saigon, Vietnam, which indicated leakage into the Saigon River [31] with approximately similar city condition cases.

2.2 Data Used in this Study

Referring to Fig. 1, some data were used to define the possible plastic waste leakage, topographical analysis, and distribution of plastic wastes' potential area of generation. To comply with the data for leakage potential, data were used in a range of statistics. Statistic data of waste management were used by two governmental hierarchies, (e.g., national and province levels). The ministry occupied the national level and national statistics bureau (BPS - Statistics Indonesia). The province level was similarly under the environmental services of Jakarta area, where it was considered on the provincal level. Both levels of data were available on their official platform websites. Some of the waste data obtained were: waste generation [32], overall waste management in each administrative city [32, 33], waste composition [32, 34], and waste treatment facilities address [32, 34].

Spatial data were used to comply with the location domain-based and topographical analysis. Spatial data used in this study consulted from several sources: the official government, NGOs, and institutions; all data were open-source and available at no cost. Non-governmental organizations also provided data in collaboration with the government. Some data were provided mainly by special divisions (e.g., slum area assessments [33, 35]) in the collaboration with the special division of civilization, KOTAKU, and the statistical bureau of Jakarta. Other data were obtained from a special division of the SMART infrastructure, where flood gates and pumps in Jakarta rivers were located [36].

Some of the spatial data used were delivered from remote-sensing products provided by the institutions. The population density was located based on the human activity, level of intensity of nightlight, which was combined with machine learning [37]; satellite-driven images of rainfall record [38]; and the Digital Elevation Model (DEM). For DEM, it was the deliverable from the National Mapping Agency of Indonesia [39], which combines multiple active remote-sensing satellites of SAR to obtain the high-resolution elevation model in meters.

2.3 Phase 1: Leakage Source Analysis

To identify the possible leakage from the plastic waste value chain, the plastic material was investigated through the material flow analysis from its generation source. The material flow of plastic waste generation was defined initially by extracting the population density at the riverbank, where humans were the source of plastic consumption. Both distribution locations were limited only to riverbanks, which was the river's buffer area within 300 m, to limit the quantification. After the sources were identified, data statistics of overall management were used to calculate the amount of plastic waste treated in waste handling and waste reduction facilities. The ratio of each plastic waste fraction was used from the ratio based on the overall management in the whole Jakarta area.

To identify possible leakage, the residential area was identified by the plastic waste generation percentage from overall composition and classifying the residential type. Slum areas [35] were provided to ease the identification of plastic waste leakage and were assigned in the vector-based data type in area definition per wards. Analysis and assessment of the residential type classification also was made by identifying each detail of land use deliverables to infer the transport mechanism of waste management. For example, high-density occupancy, such as boarding houses, remote detached houses, or stack houses, barely manage the waste, compared with the housing complex or flat. Therefore, the judgement was assessed based on the general description of the type of residence itself.

From the waste treatment facilities, leakage was detected by creating an index assessment that comprise a four-point assessments based on Minister of Public Work Regulation of Indonesia No. 3 Year 2013 [40]. They are: (1) area of the facility; (2) capacity (in ton); (3) condition of facility, including structure of the facility, segregation availability; and (4) assembled schedule. Each point of assessment was provided with score, where higher scores indicated the higher chance of leakage. Still, not all information was obtained to inform the assessment. The other alternative was to use the data of input and manage material, which kept on calculating the ratio based on the overall plastic waste management.

Finally, to quantify each source's leakage contribution to the river, the study used proximity analysis, where the closer the source of leakage with the river could bring elevated potential. In the proximity analysis, the study was marked into three zones per 100 m of riverbank. Quantification was made by using a percentage of possible plastic waste put into the river [26], which is shown in Table 1.

2.4 Phase 2: Morphometric Analysis

Morphometric analysis was chosen to create the qualified identification, a quantitative measurement approach using DEM [41], as the extracted data to deliver morphometric parameters of a watershed area [42]. To obtain its parameter, DEM are extracted to delineate the stream network (up to level 5 of it) and the watershed area, which also was generated by identification of the pour point from a well-drained basin [42]. Subsequently, the stream pathway was detected by identifying the flow accumulation, followed by calculating the stream order. In this study, Strahler's ordering system used the stream order classification method, which enabled identifying the first order to the continuing number based on its branches [43]. Then the

Table 1 Proximity analysis class quantification towards plastic waste leakage

Zone	Classification	Distance to river (m)	Percentage contribution
3	High	0–100	45%
2	Mid	≥100–200	30%
1	Low	≥200–300	15%

stream order was limited to the level 5 branch of the stream to identify the pour points. Therefore, pour points and flow direction were used to identify the water-shed area with the downstream information based on its direction. Analysis detected approximately 73 small watershed areas in Jakarta.

After delineating the watershed area, morphometric analysis could be generated by calculating the morphometric parameters. Some parameters were provided with mathematical equations towards DEM, including scale, topographic, shape, and drainage network of each watershed of that occupied river and certain canals. In this study, 18 parameters were provided to identify the runoff characteristic of each watershed that was chosen based on certain objectives to explain its characteristics shown in Table 2.

There are four adapted parameters for morphometric analysis: scale, topographic, shape, and drainage network. The scale of the watershed was identified by measur-ing the area, perimeter, and length of each watershed that positively correlated with stream discharge. Scale parameter also deemed by the calculation of time of con-centration from each watershed where it is used to explain the time estimation of runoff into the main stream. Topographic features in the watershed included the area of riverbanks, which consissted of the elevation, ruggedness, slope, and relief ratio. Elevation is needed to identify the differences of topographic features in

Table 2 Morphometric parameters

No	Type	Parameter	Unit	Range	Correlation with peak runoff*
1.	Scale	Area	km²	0.2 ha – 40 km²	Positive
2.		Perimeter	km	0.2 – 59	Positive
3.		Length	km	2.1 – 46.1	Positive
4.		Time concentration		0.034 – 1436.33	Negative
5.	Topography	Slope	o	0.00033 – 46.37	Negative
6.		Relief ratio	–	010.7	Positive
7.		Mean elevation	m	−2.12 – 53.48	Positive
8.		Ruggedness number	m	0.15 – 0.81	Positive
9.	Shape	Form factor	–	$2.1 - 10^{-6} - 1.59$	Positive
10.		Circularity ratio	–	0.065 – 0.59	Positive
11.		Compactness ratio	–	0.008 – 332.86	Negative
12.		Elongation ratio		0.0016 – 1.42	Negative
13.	Drainage network	Stream number	–	8 – 13,896	Positive
14.		Stream length	km	0.088 – 96.9	Positive
15.		Main stream length	km	1.4 – 96.9	Positive
16.		Drainage density	km¹	0.0163 – 40.7	Positive
17.		Stream frequency	km⁻²	4.72 – 11,973.6	Positive
18.		Texture ratio	km¹	1.02 – 380.32	Positive

quantification, along with relief and ruggedness that identify the surface of the watershed area. Slope is used to determine the possible velocity of the runoff. After scale and topographic parameters were calculated, they were assigned to the classification of low, mid, and high runoff class.

The third type of parameter designated to shape in quantitative measurement identifies the stream's volume and velocity. In this study, shape was defined by the form factor, compactness ratio—used to create the characteristics of the runoff intensity. Circularity ratio was used to define the catchment in the main stream and the elongation ratio was used to create the shape ratio between area and maximum length of the watershed catchment.

To identify the runoff in particular, drainage network parameters were delivered by initially identifying the stream order, which was generated in the primary stage of watershed delineation. The delivered parameters in drainage network were used to identify some points for the runoff: define the density of each watershed area, calculate ratio along with scale parameters, and characterise the stream system in various watersheds; in summary, the drainage network and its derivative parameters.

All the morphometric parameters were calculated and scored based on its class; where low-class scored 1, mid-class scored 2, and high-class scored 3. The combined class resulted in a score between 18–54, then classified into overall low-, mid-, and high-class by creating three zones based on the histogram distribution. Therefore, classification of watersheds was obtained and calculated with its plastic waste leakage contribution. To quantify the combination with stream classification, the study used percentage value of probability of plastic waste material that remained in the stream with the assumption of the lower class of runoff resulting a higher percentage of plastic waste in the river and a higher class resulting in a greater increase in the estuary. For the final calculation, accumulation in the estuary was calculated based on the main stream connected with the river outlet directly into Jakarta Bay.

2.5   Phase 3: Scenario Analysis

After plastic waste leakage and its accumulation were identified, the monthly trend of the leakage was used applying rainfall rate record. In this phase, understanding of leakage was developed with seasonal conditions and social aspects contribution. Two methods were employed by using the extreme rainfall rate (i.e., low and high rainfall), possibly bringing flood potential and long drought. The other method was developed by identifying the location of regular large, regular social events in Jakarta that were held annually.

Seasonal data were used through rainfall rate patterns during the previous 4 years (2017–now) based on the monthly average, which was driven from satellite-based precipitation CHIRPS V2.0 [44]. For the one-year recent rainfall rate, Jakarta had the lowest amount in 10 mm and 544 mm for the highest rainfall rate within

2020–2021 [44]. Patterns of 4-year rainfall rate were analyzed and classified into three classes (high, mid, low).

To create the three-zone rainfall rate, rainfall record were calculated for its monthly average and standard deviation, then defined by each month's data trend. Data is classified into three: low rainfall rate (RR >175.73 mm), mid rainfall rate (RR < 239.22 mm), and high rainfall rate (RR ≥ 239.22 mm). Therefore, the assumption developed for rainfall rate quantification, where low-class was assigned 10% in the streamflow, mid-class was assigned 50%, and high-class was assigned 75%. With the quantification, the plastic waste contribution is calculated by a formula with morphometric class and plastic waste leakage flow on an annual basis, developed by following Eq. 1 [45]:

$$PWL_{\text{Monthly}} = \left(\frac{\text{£PWL}}{12} \right) \times \%SP \times \left(\frac{MP}{3} \right)$$

(1)

where plastic waste leakage is on a monthly basis; plastic waste leakage is (PWL_{Monthly}), calculated first by equal amounts for 12 months; ΣPWL indicates the total amount of plastic waste leakage in 1 year, which was calculated in phase 2 of this study. To emphasize the rainfall rate contribution and availability of runoff, the study used the assumption of %SP, which indicated the assigned percentage for each rainfall rate class. The morphometric class, or MP was used to define the runoff category to respond to the rainfall discharge, which was valued at 1–3 according to the classified runoff in Phase 2. Therefore, the plastic waste flow into the river within months could be obtained.

In Eq. 1, the compliance for the morphometric analysis result was emphasized to create a direct information and correlation between rainfall rate trend and streamflow classification, where it is mentioned as %SP. It also defined the contribution of the subbasin analysis through watershed type. As stated in Table 2, the impact of shape and scale of the watershed included topographical features and how the condition of the drainage network impacted the response through the water input flow. This also indicates the relationship between the rainfall rate and the river area, which also brought with it the plastic waste material.

Flooding events also were considered a major part, where flood created the strong correlation with rainfall rate and caused hazardous conditions [46]. In this study, the flooding record was provided based on the most recent evidence of flooding. Compliance with historical flooding also were proven, where a 10-year trend of flooding events clearly correlated with plastic waste transport in the waterbodies [47]. Therefore, data provided by the officials for disaster and management showed the occurrence of flooding events generally took place in February. This flood pattern implied the effect from the higher rainfall rate, which frequently occurred in January and February.

Analysis of the higher rainfall, which led to the flooding record, was established to look for the possible increasing rates of plastic waste accumulation. From the data provided [48], flood hotspots were identified across the Jakarta area in February.

By using the identification with proximity analysis to calculate the distance from the river network, flood hotspots were detected in a range from 0 (right away in the river) to 2073.4 m away from the river. Quantification was made based on that distance, which classified the contribution from 5–45% impacted the runoff calculation. The quantitative value identified closer to the flood hotspot, the higher the impact to the runoff flow with regard to the higher amounts at the accumulation points.

Analysis for the social aspect contribution scenario used the annual timeline of the regular basis of large events in the Jakarta area. On a regular basis here stood for the huge event's occurrence—a previously scheduled concert held every year. There are three common events in Jakarta included in this scenario (i.e., Jakarta Fair, New Year's Eve, and Music Festival). Jakarta Fair is an annual event for celebrating the anniversary of Jakarta Province, which is held in May to June within 40 days. For New Year's Eve, the study was divided the events into three separate locations based on the type of the festival, i.e., car free night along the road with food stall provided, fireworks along with music festival. The last event is the annual Music Festival which happens regularly in December or last days of the year in 3 day big event. Each of the events documented number visited data, where Jakarta Fair received 6.8 million visitors, New Year's Eve attracted 50,000–62,000 persons, depending on the location. The Music Festival projected ot was visited by 90,000 people. Therefore, the potential leakage was calculated by using plastic waste per capita and proximity analysis zone from the location of the events to the nearest waterbodies. The study also assessed the type of the events, whether held indoors or outdoors.

3   Results

3.1   Land-Based Input Leakage Modelling

According to population density and distribution [37] and location of settlements from land use data [49], the number of people who lived within 300 m of the riverbank was 2,080,076 in population. From 45,495 residents on the riverbank, waste generation was around 221,274 tons/year, exceeding 6571.85 tons/year of plastic waste. Therefore, this amount is regarded as the baseline number of plastic waste leakage flow analysis from the first method of the residential area's generation of leakage fraction.

Waste treatment facilities were limited only to the riverbank, yet the waste could be received from other parts outside the riverbank area. Therefore, the baseline data for the overall waste generation consists of 3,048,483 tons/year waste. Plastic waste contributed 2.97% of the overall waste composition in Jakarta [32]; approximately 90,539.94 tons of plastic waste were generated. By a data input record from waste treatment in facilities, treated and managed plastic waste was estimated at 75,967.18 tons/year in the riverbank area or 83.9% input to both waste handling and waste reduction facilities there. In short, the second method, with waste treatment facilities considered, the waste generation from the overall area is the baseline number of material flow analysis.

For identification of leakage distribution from waste treatment facilities, indexing of each facility, including four categories of assessment, results in around a 0–30 index number. The assessment suggests that a higher number index indicates higher potential of leakage because of its impropriety to the standard. From overall facilities on the riverbank, 818 of 1043 waste treatment facilities deemed no leakage as a result of the compliance of the facility's conditions according to the indexing assessment method, compared with the regulation standard [40]. Results show 49,720 tons of plastic waste were not managed and possibly leaked into the river, where there also was 20,515 tons from waste-handling facilities and 29,205 tons from waste-reduction facilities. Therefore, the distribution calculated from the percentage of each index, with the distribution of possible leakage source is shown in Fig. 2. High leakage was exceedingly possible in the Central Jakarta area because of the high data input recorded.

Based on the plastic waste generated from the residential area, potential leakage distribution is calculated with the proportion ratio number of waste management records in the overall Jakarta area. Thus, 6571.85 tons/year of plastic waste are generated from the residential area on the riverbank, 98.16% are managed in the

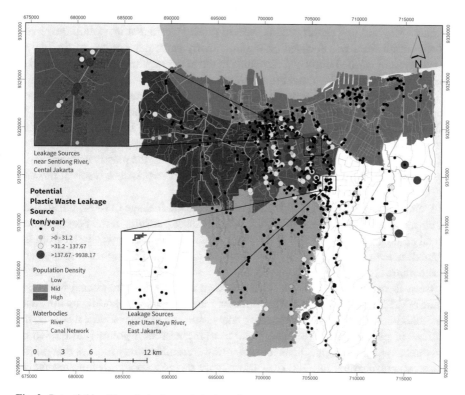

Fig. 2 Potential land-based plastic waste leakage input

waste handling and reducing facilities, taking into account the same percentage of managed plastic waste in Jakarta.

The map of Fig. 2 shows that accumulated leakage in the waste treatment facilities presents a point for each location of facilities based on data plotting. Each leakage source with various methods presents possible leakage from residential areas according to the picture element of population density in 30 m. As seen on the map Fig. 2, the smallest (black points) indicate that the waste treatment facilities emitted 0 outflow because of the same input and managed data, which mostly came from waste reserve facilities and informal sectors (one of the methods of waste reduction facilities in Jakarta). This also indicated that waste bank was an efficient method for waste reduction in Jakarta in 2020 and has a decent prospect for solid waste management [50].

3.2   Plastic Outflow into Waterbodies

With the information of proximity analysis from plastic leakage flow and runoff classification of morphometric analysis, plastic waste in the river was calculated. Sequentially, classified runoff contributes to the plastic waste outflow into the waterbodies; high runoff has a probability of retaining the material 50%, mid runoff retains 75%, and low runoff retains 90% of plastic waste that leaks into the rivers. The quantification was focused on the how many percentage points of the plastic waste was not washed away and remained in the network, continuously preferred to the connecting network as was the shifting material.

For quantification of river networks not included in the watershed, the worst case was considered, where the stagnancy of input and unknown flow overlaid it. Therefore, the unknown class of runoff was assumed to be 100% of the input that remained in the river and canal network. As seen on the map, most of the southern part of Jakarta has a red color demarcation, which implies the plastic waste material was considered as not being washed into other river networks as a result of missing information about its runoff.

Comparing the runoff classification and the plastic leakage outflow, the shifting of each watershed boundary (from the several classes) results in the possible additional amount of plastic waste in the river. Most of the shifting was input from the canals' network (presented in the short description), representing the plastic waste input.

Seen form the trend flow of plastic leakage into the river and canal networks, some of the factors may lead toward disposal into the river. Elaborating on this, with the indexing assessment and proximity analysis in the leakage source analysis, the outflow of plastic waste also tends to be from higher runoff points [28] caused by the direct vanishing of plastic as the water flows. Another study implies that plastic waste was discharged into the river profoundly owing to poor waste management procedures in the area [51]. Cities with pavements also play an important role, creating the easy drainage of water including untreated plastics because the Jakarta area has this feature. Coastal cities also take part in releasing plastics as a result of being a short distance from the rivers' area.

Through the plastic waste outflow into the river, an estimation of accumulation in the estuary was obtained by calculating the flow input into the main stream, as presented in Fig. 3. Compared with another similar study, view the checking and overlapping of the coastline data; 7 points of estuary leading directly into Jakarta Bay were generated. From the calculation, accumulation results in the range of plastic waste leakage to the ocean range from 23.92–5973.97 tons/year, which varies based on the main stream's flow, slum area detection, and strain from floodgates or pumps in the waterbodies.

The highest accumulation in the eastern estuary (EP5) in the Sunter River exceeds 5973.97 tons/year. This was followed closely by the estuary point from Sunter; EP6 accumulates 3412.69 tons/year of plastic waste leakage. The main one in Jakarta, the Ciliwung River, exceeds 1245.6 tons/year of plastic waste leakage, surprisingly not the highest leakage source as expected. The lowest accumulation emitted by the western part of Jakarta into the Cengkareng Drain was 23.93 tons/year of leakage. The overall results and its data that were delivered to the main stream are shown in Table 3.

As mentioned already, variation of the accumulation in 1 year could be differentiated by the main stream flow, slum area detection, and strain from floodgates or pumps into the waterbodies. As the addition brought from the main stream calculation, the constraint of the calculation was the increase based only from the final calculation in the main stream. In this study, the assumption was made by the analysis of runoff flow and included the inputs from canals that resulted in the flow into the main stream. Elaborating, with the streamflow classification from each main stream, runoff was greatly impacted by the buildup number, where the disposal of plastic waste outflow was retained if the leakage input was to the lower classified runoff, as seen from the results shown in Table 3. Therefore, the assumption in flow quantification, by using the hierarchy based on the percentage to differentiate the class, could be used.

Accumulation in the estuary also was strained by the slum area distribution, where slums indicate the direct disposal practices and low administration of the area's waste management. Because the distribution, as shown in Fig. 3, in a slum area commonly is spread in the ward located at the coastal area, which also is close to the estuary location. Therefore, the direct disposal impacts the constrained accumulation at some points, namely EP3, EP4, EP5, EP6, and EP7. From the initial result of Fig. 3, the location of the estuary in EP5 and EP6 was very dense with a possible outflow source. This implied that the greater estimation takes place in EP5 and EP6 locations. On the other hand, EP5 has more branches from the canals, even though the streamflow was classified as mid-class runoff. These branches create the advanced possibility of plastic waste transport into the waterway.

Another constraint for accumulation from the main stream also was impacted by the utilization of floodgates and pumps in the river. A close approach of floodgate location in the main stream situated close to the estuary area leads to the smaller amount of plastic waste outflow reaching the estuary zone. As shown in Table 3,

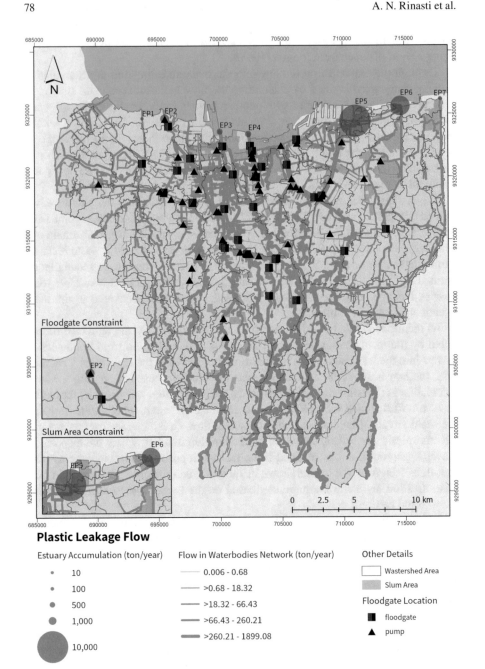

Fig. 3 Accumulation and pathway of plastic waste outflow

Table 3 Estimation of the leakage accumulation

Point no.[a]	Mainstream name	Streamflow classification	Affected by floodgate	PW accumulation (tons/year)
EP1	Cengkareng	Low	Yes	0.239×10^2
EP2	Angke–Ciliwung	High	Yes	12.45×10^2
EP3	Opak	High	Yes	6.27×10^2
EP4	Ciliwung Gunung Sahari	Mid	Yes	6.61×10^2
EP5	Sunter	Mid	Yes	59.73×10^2
EP6	Cakung	High	No	34.12×10^2
EP7	Kanal Timur	Low	No	1.97×10^2
Total				**12,142.21**

[a]Order based on the estuary position from west to east

some estuary locations that were affected by the floodgates ensured reduced plastic waste generation. Highlighted in EP2, where the discharge of the plastic waste from Ciliwung River, floodgates and pumps also reveal the decreased uptake of buildup. As in the main river, rejuvenation in the Ciliwung River also impacted the accumulation in EP2, which resulted in less plastic waste being emitted into this location compared with EP5 and EP6.

3.3 Monthly Trends and Scenario Applied

By using the satellite-based rainfall record [44], where the Jakarta area covered approximately 42 pixels of the rainfall image, the quantification of streamflow becomes varied and leads to the difference in monthly plastic waste outflow into the river. Starting from month of May, which also profoundly is the transition month from rainy to dry season in most years in Indonesia.

Decline occurs for plastic waste outflow from February to March and is slightly lower in April, where its maximum value in March drops to 15.82 tons/month. During the period of March and April, plastic waste emitted into the river was reduced. However, some river networks in the southern part of Jakarta remain high in discharge, ranging from 5.58–11.81 tons until April. This indicates the sounder correlation of rainfall to leakage, where the average rainfall rate occurs as 151.63 mm. It also suggested the accumulation in the estuary in which the rainy season greatly impacted discharge of more plastics.

Analysis in the various conditions was applied to understand how the actual condition and the model can be used for several scenarios. In terms of rainfall rate as the indicator for different trends per month for the accumulation in the estuary, flooding events—in regard to long drought possibilities—and regular social events simulated release into the river networks. Both major events impacted the increase of plastic waste leakage, which is shown in the data trend quantitively in Fig. 4.

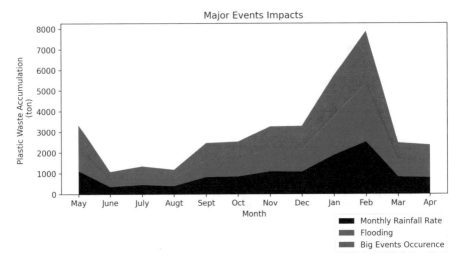

Fig. 4 Monthly trend of plastic waste leakage accumulation from various major scenarios

The monthly average rainfall rate produced from the calculation of plastic waste outflow time, the weight calculation per month was obtained, and total accumulation results(see Fig. 4) for the first trend (colored black). As expected in the time series map analysis, February reaches the highest accumulation in total into Jakarta Bay by 2516.44 tons/month. The lowest accumulation collects in June when plastic waste leaks approximately 353.39 tons/month. Based on the extreme peak value, the rainfall rate was impacted fully to the accumulated total per month, followed by a weight calculation, where February reaches 0.21 out of total of 1. This explains that, according to 1 year plastic waste leakage, almost one-quarter of the annual total was emitted during February because of its high rainfall rate starting from January. It also was proven that rainfall calculation in the river brought the pulses of plastic waste release from the river into the sea [52].

From the first scenario, a study was developed by the flooding record and overlaid with the rainfall calculation. By using the recent flooding record and identifying 386 flooding hotspots that occurred in February, the impact is shown in Fig. 4—that is, flood increases the plastic waste accumulation by 321.5 tons or increases by 12.78% compared to the monthly average rainfall rate. Based on the flooding scenario, there are some effects to be concerned about regarding the rejuvenation of the river, which also brought the lower amount of emissions during the flooding events. For anticipating the flood, the Government of Jakarta prepared a greater intensity of river cleaning on the starting point of rainfall increases (profoundly by November) by Pasukan Orange—a special unit from Environmental Services of Jakarta [28]. Because the flooding events in Jakarta are mostly flash floods and river floods, usually because of the overflow from the other neighboring highland cities, the government instituted the rejuvenation, especially in the main streams in the Jakarta area (i.e., Cilliwung River). Another concern, also impacted by the floodgates, was to prevent the greater outflow, in which more possibilities of plastic waste leakage were blocked from going into the connected canals with floodgates.

The other scenario developed was in accordance with the occurrence of regular large social events, which emphasizes how the concentrated gathering of numerous people at one time could affect plastic waste emission into the river. The main event is the Jakarta Fair, which brings together the most people in June for 40 days straight at one venue at Central Jakarta and located 228.47 m fron the nearest river (i.e., West Pademangan River). With a total of millions gathered for the cumulative number of days for the event, forecasts indicate the Jakarta Fair contributes about 13 tons, or an increase of 3.86%, into the river in June. Although predictable the 1972 tons of plastic waste could be generated based on plastics consumption per capita, the outflow might be able to be managed because of the type of event. Jakarta Fair is well-managed and the officials are liable so proper execution might be achieved.

The second designated event is New Year's Eve, where the celebration could be anywhere all over Jakarta. By focusing on three different locations—Ancol, Bundaran HI, and Taman Mini, Indonesia Indah (TMII)—results are expected to be around 97.52 tons of plastic (contributing an increase of plastic waste leakage of 5.2%); this could be emitted into the river in January after the event. One of the considerations for this high amount of plastic waste leakage is the high density of visitors, which also supplemented by the food stalls available at the events. Similar to the common entertaining events, most food stalls were using single-use plastic containers. In addition, most of the event was held outdoors, which implies the greater possibility of leakage into the river. Furthermore, New Year's Eve in Bundaran HI was placed along the road which was approximately 247 m from canals that connected directly to the Cideng River. From the river network, it is highly possible to impact the EP3 estuary point from Opak River's main stream.

Another big event designated is the Music Festival which usually is held in mid-December for 3 days. With the same location as the Jakarta Fair, this event forecasted a contribution of about 65.01 tons, or 6.07% increase for plastic waste leakage. Both New Year's Eve and the Music Festival are considered by the timing where it was held in the usual rainy season in Indonesia with an increased rainfall rate (compared to when the Jakarta Fair is held). In conclusion, big regular events which held annually increase the plastic waste leakage in regard to the months of the event.

4 Discussions

4.1 Plastic Material Pathway from Hydrological Response

From the leakage source input into the river, plastic waste hotspots within the river flow was amplified by using the topographic features and subbasin analysis. Thoughtfully, the morphometric analysis that brought the mathematical parameters for subbasin analysis include the topography, resulting in three classes of possible runoff based on the elevation model. Figure 5 shows that there are 9 of 73 delineated watershed categorized as high runoff, which means these watersheds are very potentially able to bring the plastic material into the end buildup point in the estuary and contribute less density to the river. It also was regarded to be that the higher

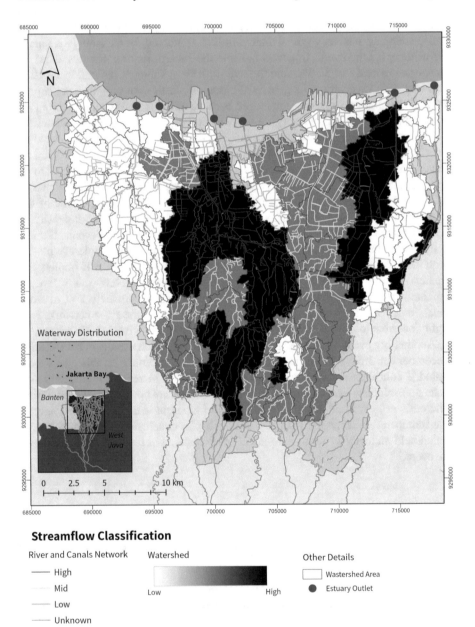

Fig. 5 Morphometric analysis results for streamflow classification

discharge into the river leads to the velocity of river flow through plastic material to be carried along its pathway [31]; yet, the abundance of plastic waste was affected differentially with the streamflow classification. Mid-class classified runoff emphasized the moderate stage of runoff on average and had less significance in transporting material than the high runoff, leading to the low-class runoff with a low significance of transport into the river network.

Based on the morphometric parameters generated from DEM, some larger watershed areas, mainly with high runoff as the parametric scale led to higher recipients of water from other areas [42, 53], which also defined some main streams in the larger watershed employed higher runoff into the estuary. One of the highly significant for the topographical features included the elevation, slope, ruggedness, and relief; where a steeper slope results in greater streamflow. It also indicated where the main streams in the Jakarta area have significant topographical features, implied by the higher runoff into the river. Some parts of the big river in Jakarta also comply with a more complex drainage network, resulting in the high-class impact from the drainage network parameters.

The results shown in Fig. 5 indicates that watersheds were classified into several classes: high, mid, low, and unknown. This judgement for each class identified how the response of each watershed and waterbodies network delineation to the input of water. In this study, the result shows that the classification response to the streamflow led to the water flow as the medium of plastic waste transport in the river pathway. High class represents the plastic waste carried more to the accumulation point. Differentiated from the lower classes (i.e., mid and low), represents that the plastic material accumulates within the flow. Therefore, the concept of the morphometric class shows how plastic materials possibly accumulate or become inputs into the other network of water bodies.

According to Fig. 5, some river networks have an unknown class of streamflow. The map shows that some connected rivers to the outside of the Jakarta area are not included and obtained the classification of unknown river networks. In this case, watershed delineation was correlated with pour points identified. Pour point identification in this study used the principal of network with 5 stream levels [42], occasionally some of the unrelated network could be denied. It also wasdetermined by the network delineation form, DEM, for which there are some possibilities of unrealistic river networks (compared with the polyline feature from the officials). Because of the limited pour point selected, some areas were not included—mainly in the coastal areas and the suburbs. In this case, pour points were limited and not included in the estuary points nor river outlet, which affected the unidentified watershed in the coastal area. For suburbs with unknown value, it was specified that these areas are included in the other watershed outside the administrative Jakarta area. This also was improved from previous research where two watersheds were clearly not included in the study area—located in Tangerang in the western part and Bekasi in the eastern part [28].

For river and canal networks that were connected and included in one watershed, its runoff type instantly classified the same as the watershed class. As seen in Fig. 5, the high runoff of the watershed area affects the river streams and canal networks, making them high-class of runoffs. It also presented the higher river runoff regardless in the central part of Jakarta, where the main river of Jakarta (i.e., Ciliwung River) is located in the high runoff watershed. Ciliwung River's high runoff also impacted its pathway across the cities and its upstream based on the highland in Bogor Regency (southern Jakarta boundaries). High runoff also impacted the rainfall [42], where the Ciliwung River frequently impacted flooding events in Jakarta [48], by way of other evidence that a higher class of runoff will be extremely affected by meteorological conditions.

4.2 Leakage Comparisons

To make the estimation in accumulation concise, the results were compared with the other studies in the previous year, starting from the global model of plastic waste input in 2015 [7], monthly variation of survey-based accumulation in Jakarta Bay during the period 2015–2016 [28], and visual counting combined with the hydrodynamic model in the Jakarta area during the period 2018–2019 [29]. Compared with the location, there are 4 estuary spots with a global model, 7 estuary spots with survey-based accumulation, and 5 estuary spots with the latest model. Comparison was made by plotting the results with the representative previous model, which is shown in Fig. 6.

In general, the results using the recent waste management data has a higher estimation than the previous year. According to the whole previous year's model, results

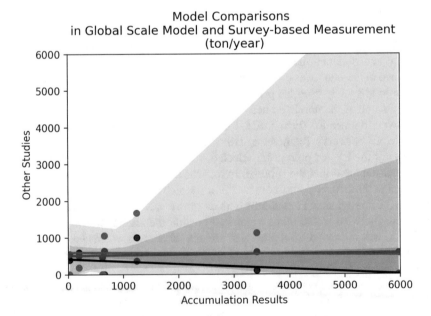

Fig. 6 Comparisons with other studies at the same location throughout the years

were exceeding high in EP5 and EP6. EP5 only compared with the survey-based model where the amounts of plastic waste are 10 times greater than the measurement in the previous 6 years. EP6 had the closest amount of leakage with the global model in 2015 and a higher gap with the latest model, exceeding 34 times more. Although there were some underestimated accumulations in EP1, which only was compared with the prior year's model and the survey-based model; the result only was able to record 5.98% and 4.37%, respectively, from the previous year's studies.

Some possibilities could be analyzed—that is, where the recent study used the most current waste management data with some constraints and a different approach to the study occupancy. In this study, the waste management data used started from 2020–2021, whereas the matter-of-fact year 2020 was the urgent special case owing to the pandemic that happened. A metropolitan area, like Jakarta, with homes for the population of 10,534,339 there, have a higher chance of increasing waste generation started in year 2020 because of the intensity of human activity with the new adaptations of consumer behaviour as a result of the pandemic. The study by LIPI, an official science and research institution in Indonesia, implied that Jakarta's capital city has increased its own plastic waste generation caused by the sudden-rise of single-use consumption behaviour [54].

5 Conclusions

Contextualizing with providing the early warning system for plastic waste leakage with topographical concerned, modelling in forecasting the amount of plastic waste leakage in the river through the estuary is applicable for the city level. With some requirements of data: waste management statistics in city level from generation to management input, digital elevation model, and rainfall rate. With these following data, model of plastic waste leakage could be implemented and utilized in the other cities in Indonesia.

Regarding to the result of model in Jakarta, estimation of leakage is applicable to define the plastic waste disposal trends, as some important points:

1. Identification of plastic waste in the land detected by identify the properness of the waste treatment facilities, proximity analysis and residential type identification; results the maximum leakage 9938.17 ton in annual basis. Estimating the input might vary due to the location of source to the river which possibly leaked to 24.14 ton/year.
2. Plastic waste leakage in the river detected by improvising the hydrological characterization in the river networks, accumulated the plastic waste by 1899.08 ton/year in the mainstreams.
3. Accumulation in the estuary might vary considering the blockage strain, rainfall rate and the streamflow identification, which detected the highest accumulation in February by 2837.94 ton/month plastic releases to the Jakarta Bay. By developing extreme scenarios, plastic waste is forecasted to be emitted by 12.78%.

Due to the study is compliance for baseline data, policy making could be established in Jakarta by emphasizing the mitigation in priority river with overwhelming plastic. Setting up the cleanup activity in regular basis by the community and prevention by monitoring also potentially implemented in Jakarta's River area, especially within the riverbank area. Within the hotspot, identifying plastic waste in the river is also one of the solutions to proof the management quantification for waste management.

Further improvements for mitigating the plastic leakage are also enabled with the improvement on the monitoring systems between the land-based activities and river flow including the application of machine learning. Applying the CCTV along with the machine learning algorithm can be developed to improve the trend of the leakage to the river. A contribution from citizen science also enriched the study of plastic leakage behavior, building awareness towards the society.

References

1. Geyer, R., Jambeck, J.R., Law, K.L.: Production, use, and fate of all plastics ever made. Sci. Adv. **3**, 1–5 (2017)
2. Cózar, A., Echevarría, F., González-Gordillo, J.I., Irigoien, X., Úbeda, B., Hernández-León, S., Palma, Á.T., Navarro, S., García-de-Lomas, J., Ruiz, A., Fernández-de-Puelles, M.L., Duarte, C.M.: Plastic debris in the open ocean. Proc. Natl. Acad. Sci. U.S.A. **111**(28), 10239–10244 (2014)
3. Ncube, L.K., Ude, A.U., Ogunmuyiwa, E.N., Zulkifli, R., Beas, I.N.: An overview of plastic waste generation and management in food packaging industries. Recycling. **6**(12), 1–25 (2021)
4. Galgani, L., Beiras, R., Galgani, F., Panti, C., Borja, A.: Editorial: impacts of marine litter. Front. Mar. Sci. **6**(208), 4–7 (2019)
5. Ryan, P.: The transport and fate of marine plastics in South Africa and adjacent oceans. Marin. Plast. Debris Rev. Art. **116**(5), 1–9 (2020)
6. Verster, C., Bouwman, H.: Land-based sources and pathways of marine plastics in a South African context. Marin. Plast. Debris Rev. Art. **116**(5/6), 1–9 (2020)
7. Lebreton, L.C.M., Van Der Zwet, J., Damsteeg, J.W., Slat, B., Andrady, A., Reisser, J.: River plastic emissions to the world's oceans. Nat. Commun. **8**, 1–10 (2017)
8. Forbes, E., Mccauley, D., Kordell, T., Joyce, F., Visalli, M., Morse, M.: River plastic pollution: considerations for addressing the leading source of marine debris. Nature, 1–28 (2019). https://doi.org/10.13140/RG.2.2.20699.80165
9. Sakti, A.D., Rinasti, A.N., Agustina, E., Diastomo, H., Muhammad, F., Anna, Z., Wikantika, K.: Multi-scenario model of plastic waste accumulation potential in indo-nesia using integrated remote sensing, statistic and socio-demographic data. ISPRS Int. J. Geo Inf. **10**(7), 1–28 (2021)
10. van Emmerik, T., Schwarz, A.: Plastic debris in rivers. Wiley Interdiscip. Rev. Water. **7**(1), 1–24 (2020)
11. Jambeck, J.R., Geyer, R., Wilcox, C., Siegler, T.R., Perryman, M., Andrady, A., Narayan, R., Law, K.L.: Plastic waste inputs from land into the ocean. Sci. Mag. **347**(6223), 768–770 (2015)
12. Eunomia: Plastics in the Marine Environment. Eunomia Research & Consulting, Bristol (2016)
13. Barnes, D.K.A., Galgani, F., Thompson, R.C., Barlaz, M.: Accumulation and fragmentation of plastic debris in global environments. Philos. Trans. R. Soc., B. **364**(1526), 1985–1998 (2009)
14. Lebreton, L., Andrady, A.: Future scenarios of global plastic waste generation and disposal. Palgrave Commun. **5**(1), 1–11 (2019)

15. Marazzi, L., Loiselle, S., Anderson, L.G., Rocliffe, S., Winton, J.: Consumer-based actions to reduce plastic pollution in rivers: a multi-criteria decision analysis approach. PLoS ONE. **15**(8), 1–15 (2020)
16. Walls, M., Palmer, K.: Upstream Pollution, Downstream Waste Disposal, and the Design of Comprehensive Environmental Policies, vol. 1, pp. 94–108 (2001)
17. Manfredi, E.C., Flury, B., Viviano, G., Thakuri, S., Khanal, S.N., Jha, P.K., Maskey, R.K., Bhakta, R., Kafle, K.R., Bhochhibhoya, S., Ghimire, N.P., Babu, B., Chaudhary, G., Giannino, F., Cartenì, F., Mazzoleni, S., Salerno, F.: Solid waste and water quality management models for Sagarmatha National Park and Buffer Zone, Nepal. Mt. Res. Dev. **30**(2), 127–142 (2010)
18. Monroe, L.. Waste in Our Waterways: Unveiling the Hidden Costs to Californians of Litter Clean-Up. NRDC Issue Brief August 2013. 1–5 (2013)
19. Ferronato, N., Torretta, V.: Waste mismanagement in developing countries: A review of global issues. Int. J. Environ. Res. Public Health. **16**, 1–28 (2019)
20. Gall, S.C., Thompson, R.C.: The impact of debris on marine life. Mar. Pollut. Bull. **92**(1–2), 170–179 (2015)
21. Law, K.L.: Plastics in the marine environment. Ann. Rev. **6**(52), 1–25 (2016)
22. Wilcox, C., Mallos, N.J., Leonard, G.H., Rodriguez, A., Denise, B.: Using expert elicitation to estimate the impacts of plastic pollution on marine wildlife. Mar. Policy. **65**, 107–114 (2016)
23. Barboza, L. G. A., Cózar, A., Gimenez, B. C. G., & Barros, T. L. Macroplastics pollution in the marine environment. In C Sheppard World seas: an environmental evaluation (2). Elsevier Ltd. (2019)
24. Vorosmarty, C.J., Fekete, B.M., Meybeck, M., Lammers, R.B.: Global system of rivers: Its role in organizing continental land mass and defining land-to-ocean linkages. Glob. Biogeochem. Cycles. **14**(2), 599–621 (2000)
25. Schmidt, C., Krauth, T., Wagner, S.: Export of plastic debris by Rivers into the sea. Environ. Sci. Technol. **51**(21), 12246–12253 (2017)
26. Shuker, I. G., & Cadman, A.: Marine debris hotspot rapid assessment: Synthesis report. Working Paper Report No. 126686, 1, 1–46 (2018)
27. Petrlik, J., Ismawati, Y., DiGangi, J., Arisandi, P., Bell, L., Beeler, B.: Plastic waste poisons Indonesia's food chain: Indonesia egg report. Toxics Free, 1–20 (2019)
28. Cordova, M.R., Nurhati, I.S.: Major sources and monthly variations in the release of land-derived marine debris from the Greater Jakarta area, Indonesia. Sci. Rep. **9**, 1–8 (2019)
29. van Emmerik, T., Loozen, M., van Oeveren, K., Buschman, F., Prinsen, G.: Riverine plastic emission from Jakarta into the ocean. Environ. Res. Lett. **14**(8), 1–10 (2019a)
30. National Plastic Action Partnership. Radically reducing plastic pollution in Indonesia: a multi-stakeholder action plan national plastic action partnership. World Economic Forum, (2020)
31. van Emmerik, T., Strady, E., Kieu-Le, T., Nguyen, L., Gratiot, N.: Seasonality of riverine macroplastic transport. Sci. Rep. **9**(1), 1–9 (2019b)
32. Kementrian Lingkungan Hidup dan Kehutanan Republik Indonesia (KLHK). (2021). Sistem Informasi Pengelolaan Sampah Nasional [Web source]. https://sipsn.menlhk.go.id/sipsn/. Accessed on 01 June 2021
33. Bureau of Statistic Center (BPS) Indonesia. (2020). Hasil Sensus Penduduk 2020 [PowerPoint slides]. https://www.bps.go.id/website/materi_ind/materiBrsInd-20210121151046.pdf
34. Environmental Services (DLH) Jakarta. (2020). Data Sampah Jakarta [Web source]. https://upst.dlh.jakarta.go.id/wastemanagement/data. Accessed on 01 June 2021
35. KOTAKU. (2020). Peta Sebaran Kumuh Provinsi DKI Jakarta [Web map portal]. https://www.arcgis.com/apps/Embed/index.html?webmap=1aa1482cb1b14afa8993ffa7d303f25c & extent=106.6613,-6.2901,107.0204,6.1027 & home=true & zoom=true & scale=true & details=true & legendlayers=true & active_panel=details & basemap_gallery=true & disable_ scroll=true & theme=light. Accessed on 01 June 2021
36. Smart Infrastructure Facilities. (2015). SMART Metadata [Web map portal]. https://smart-metadata.eis.uow.edu.au/geonetwork/srv/eng/search#|1ecb8a02-5c94-4532-ad19-ce690b8b0c21. Accessed on 01 June 2021

37. Facebook Connectivity Lab and Center for International Earth Science Information Network—CIESIN—Columbia University. (2016) High Resolution Settlement Layer (HRSL) [Web source]. https://www.ciesin.columbia.edu/data/hrsl/. Accessed on 01 June 2021
38. Climate Hazards Center UC Santa Barbara. (2021). CHIRPS: Rainfall Estimates from Rain Gauge and Satellite Observations [Web map portal]. https://www.chc.ucsb.edu/data/chirps. Accessed on 01 June 2021
39. DEMNAS National Mapping Agency Indonesia (DEMNAS BIG). (2021). Seamless Digital Elevation Model (DEM) dan Batimetri Nasional [Web map portal]. https://tanahair.indonesia.go.id/demnas/#/. Accessed on 01 June 2021
40. Regulation of Ministry of Public Works Indonesia No. 3 Year 2012 (Permen PU RI No. 3 Tahun 2012). (2013). http://ciptakarya.pu.go.id/plp/upload/peraturan/Permen _PU_No_3_ Tahun_2013_-_Penyelenggaraan_PS_Persampahan.pdf. Accessed on 01 June 2021
41. Asfaw, D., Workineh, G.: Quantitative analysis of morphometry on Ribb and Gumara watersheds: implications for soil and water conservation. Int. Soil Water Conserv. Res. **7**(2), 150–157 (2019)
42. Adnan, M.S.G., Dewan, A., Zannat, K.E., Abdullah, A.Y.M.: The use of watershed geomorphic data in flash flood susceptibility zoning: a case study of the Karnaphuli and Sangu river basins of Bangladesh. Nat. Hazards. **99**, 425–448 (2019)
43. Hughes, R.M., Kaufmann, P.R., Weber, M.H.: National and regional comparisons between Strahler order and stream size. J. N. Am. Benthol. Soc. **30**(1), 103–121 (2011)
44. Funk, C., Peterson, P., Landsfeld, M., Pedreros, D., Verdin, J., Shukla, S., Husak, G., Rowland, J., Harrison, L., Hoell, A., Michaelsen, J.: The climate hazards infrared precipitation with stations—a new environmental record for monitoring extremes. Sci. Data. **2**, 1–21 (2015)
45. Rinasti, A.N., Ibrahim, I.F., Gunasekara, K. et al. Fate identification andmanagement strategies of non-recyclable plastic waste through the integration of material flow analysis and leakage hotspotmodeling. Sci Rep 12, 16298 (2022). https://doi.org/10.1038/s41598-022-20594-w.
46. Kamal, A.S.M.M., Shamsuddduha, M., Ahmed, B., Hassan, S.M.K., Islam, M.S., Kelman, I., Fordham, M.: Resilience to flash floods in wetland communities of northeastern Bangladesh. Int. J. Disaster Risk Reduct. **31**(Dec 2017), 478–488 (2018)
47. Roebroek, C.T.J., Harrigan, S., Van Emmerik, T.H.M., Baugh, C., Eilander, D., Prudhomme, C., Pappenberger, F.: Plastic in global rivers: are floods making it worse? Environ. Res. Lett. **16**(2), 1–12 (2021)
48. Bureau of Regional Disaster Management (BPBD) Jakarta. (2021). SHP_Jakarta Feature Data [Webmap Portal]. https://jakartasatu.jakarta.go.id/server/rest/services. Accessed on 01 June 2021
49. Open Maps Jakarta Satu. (2021). Open Maps Jakarta Satu [Web map portal]. https://jakartasatu.jakarta.go.id/portal/apps/sites/?fromEdit=true#/public/pages/unduh. Accessed on 01 June 2021
50. Khair, H. (2019). Study on waste bank activities in Indonesia towards sustainable municipal solid waste management. Doctoral disertation, The University of Kitakyushu. https://kitakyu.repo.nii.ac.jp/index.php?action=pages_view_main & active_action=repository_action_common_download & item_id=714 & item_no=1 & attribute_id=20 & file_no=1 & page_id=13 & block_id=294. Accessed on 01 June 2021
51. Meijer, L.J.J., van Emmerik, T., van der Ent, R., Schmidt, C., Lebreton, L.: More than 1000 rivers account for 80% of global riverine plastic emissions into the ocean. Sci. Adv. **7**(18), 1–14 (2021)
52. Ryan, P.G., Perold, V.: Limited dispersal of riverine litter onto nearby beaches during rainfall events. Estuar. Coast. Shelf Sci. **251**, 1–9 (2021)
53. Abdel-Fattah, M., Saber, M., Kantoush, S.A., Khalil, M.F., Sumi, T., Sefelnasr, A.M.: A hydrological and geomorphometric approach to understanding the generation of wadi flash floods. Water (Switzerland). **9**(7), 1–27 (2017)
54. Setiawan, V. N., & Tobing, S. (2020, December 2). SOS Sampah Pandemi di Ibukota [Blog post]. https://katadata.co.id/arsip/analisisdata/5fc719de77307/banjir-sampah-plastik-selama-pandemi. Accessed on 01 June 2021

Measurement of PM_{10}, $PM_{2.5}$, NO_2, and SO_2 Using Sensors

Vinit Lambey and A. D. Prasad

Abstract The objective of this study is to measure a few air quality parameters such as PM_{10}, $PM_{2.5}$, NO_2, and SO_2 using sensors at the National Institute of Technology (NIT) Raipur campus. All those parameters are measured using three sets of sensors. Measurement of SO_2 has been done using SPEC DGS-SO2 and SPEC 3SP_SO2-20P which are electrochemical sensors, while the third sensor Horiba APSA 370 is based on ultraviolet fluorescent (UVF) method. Measurement of NO_2 has been done using Alphasense NO2-B43F, SPEC 3SP_NO2_5FP (both electrochemical sensors), and Horiba APNA 370 (chemiluminescence approach based) sensors. Particulate matter concentration has been monitored using Plantower PMS7003, Prana Air PAS-OUT-01 (both based on light scattering principle), and Met One Instruments BAM 1020 (based on beta attenuation method) sensors. All these sensor sets are installed at three monitoring stations (MS) namely, MS-1, MS-2, and MS-3. Current study results are compared with the previous studies. The results obtained in the current study have been compared with the previous studies. The percentage difference for $PM_{2.5}$ is found to be (−) 107.28% between MS-1 and reference study. The difference for MS-2 and MS-3 with reference study for $PM_{2.5}$ is found to be (−) 104.62% and (−) 113.74%, respectively. The difference between concentration values obtained from MS-1, MS-2, and MS-3 with reference study for PM_{10} has been observed to be (−) 124.18%, (−) 131.60%, and (−) 125.39%, respectively. The percentage difference for NO_2 between MS-1 and reference study is found to be (−) 55.90%. The difference for MS-2 and MS-3 with reference study for NO_2 is found to be (−) 66.39% and (−) 58.16%, respectively. The difference between MS-1, MS-2, and MS-3 with data obtained from reference study for SO_2 has been observed to be 14.81%, 28.59%, and (−) 10.12%, respectively. Negative value indicates that observed concentration is less than reference study data. The main reason for obtaining less concentration of air quality parameters from sensors is location of sensor installation. Due to the location, the sensors are less susceptible to the pollution caused by traffic and other anthropogenic activities. The main rea-

V. Lambey (✉) · A. D. Prasad
Civil Engineering Department, National Institute of Technology Raipur, Raipur, India

© The Author(s), under exclusive license to Springer Nature Switzerland AG 2023
W. Boonpook et al. (eds.), *Applied Geography and Geoinformatics for Sustainable Development*, Springer Geography,
https://doi.org/10.1007/978-3-031-16217-6_6

89

son for conducting current study for short period is difficulty faced during setting up and calibration of sensors of MS-1 and MS-2.

Keywords Air quality parameters · PM_{10} · $PM_{2.5}$ · NO_2 · SO_2 · Gas sensor

1 Introduction

Nowadays air quality is a concern for humanity and animals due to rapid urbanization. To counter negative impacts due to poor air quality, a prevalent air quality assessment technology can enable us to better comprehend the quality of life. By integrating data on air quality and health outcomes, empirical connections between air contaminants and human health can be discovered [19]. These empirical connections can be used to evaluate early warning limits and health hazards associated with environmental exposure. As a result, many air quality monitoring studies use precise but costly devices to measure differences in air pollution on a wide scale, such as over a vast area [10]. Majority of the data used in these researches are obtained from local authorities or research institutions that set up limited monitoring stations throughout the vast area to capture overall air pollution levels. By evaluating the general patterns of a large-scale region, high-resolution regional variation of air quality cannot be identified. The establishment of an air quality monitoring system consisting of small sensors can benefit in the investigation of interactions between air contaminants and human activities at very fine scale. High spatial-temporal fluctuations in air quality can be captured by the sensor-based monitoring system. Low-power consumption and improved measurement precision can be accomplished simultaneously with the rapid growth of the manufacturing process in semiconductor technology, such as reduced chip sizes and new sensing materials, in recent years [20]. In urban environments, it is now feasible to establish an efficient wireless sensor network for environmental monitoring and research [15].

Recent advances in sensor technology have resulted in the establishment of a new air pollution monitoring approach [14, 23]. Electrochemical sensors have been recognized as one of the most viable sensor technologies for measuring inorganic gases at sub-ppm levels. In a real-world context, optical counters have also been employed as promising low-cost sensors for particle pollution. Borrego et al. [2, 3] did a joint operation with hundred gas sensors operating side by side with reference analyzers, installed in a mobile air quality laboratory, and working for a 2-week campaign in Aveiro, Portugal, to analyze the air quality microsensors. Exceptional research in European countries such as the United Kingdom [16], Norway [5], Switzerland [11], Belgium [18], and Italy [17] demonstrated that explanatory levels of air pollutants can be detected using accurate gas sensor systems for air quality monitoring.

Wen et al. [26] developed a low-power consumption wireless sensor network for measuring CO consisting of multiple node types. The hourly variations of CO between weekdays and weekends have been observed, and comparison of the spatiotemporal patterns of air quality between rush hour and regular traffic for 1 day has been observed to capture possible potential risk of human exposure to traffic-related air pollution. Devarakonda et al. [9] created a mobile device that can calculate real-time CO levels of the area, the node of which consists of a wireless sensor connected to a smart phone that serves as an interface. Rohi et al. [21] determined the concentration of O_3, CO, NO_2, CO_2, SO_2, NH_3, and PM emissions using an airborne platform. Hu et al. [13] developed an Internet of Things (IoT)-based 3D air quality monitoring system, and the obtained data has been processed using spatial fitting technique. In another study carried out by Carrozzo et al. [4], multi-sensor node has been used to measure the CO, NO_2, and O_3 concentration from smoke generated by brushwood-operated controlled fire. No standard equipment has been used for validation. Bing et al. [1] measured the CO using metal oxide semiconductor gas sensor. The 3D kernel DM algorithm was used to build a three-dimensional distribution map of the CO gas. Thu et al. [24] describe an air quality monitoring system that uses IoT sensors and long-range communication. It includes an Arduino microcontroller with a long-range module as well as a sensor kit for monitoring temperature, humidity, dust, and CO_2. It's vital to note that while the study used a sensor capable of measuring dust particle concentrations, it doesn't specify $PM_{2.5}$ or PM_{10} concentrations, instead assessing the quantity of solid particles in the air in general. Kadri et al. [13] built a WSN for exterior air pollution observation, and the developed model was implemented in real time. The sensors collect pollutant data such as O_3, NO_2, CO, and H_2S, which is then sent to a server through GPRS. The sensor nodes that are stationary are powered by a solar module. Air pollution data is made available to the public through specialized mobile and online apps. Desai and Alex [7] created an air pollution monitoring and forecasting system for CO and H_2. Their solution includes a BeagleBone Black microcontroller, MOS (MQ-7, MQ-11) sensors, and a GPS module for monitoring pollution levels. The information gathered by the sensors is uploaded to Microsoft Azure using Python SQL. They used a machine learning service to anticipate pollution using data stored in the cloud. This research shows how cloud data can be utilized to forecast air pollution.

This study has been conducted in National Institute of Technology (NIT) Raipur campus. As Raipur is one of the most polluted cities in India, study area is chosen to monitor particulate matter ($PM_{2.5}$, PM_{10}), nitrogen dioxide (NO_2), and sulfur dioxide (SO_2) using three sets of sensors. Monitoring using the sensors in the study area has not been attempted in the previous research.

2 Materials and Methods

2.1 Study Area

NIT Raipur campus is chosen as study area. The monitoring stations have been set up at the roof top of administrative block in the campus. The study location is about 200 m away from the major traffic road. Figure 1 shows the location of monitoring

Fig. 1 Study area

stations with star symbology. The coordinates of the three monitoring stations are
21°15′00.29″ N and 81°36′16.63″ E. Study area has been observing an average
annual rainfall of around 1400 mm. It has tropical wet and dry climate. The mean
temperature in the region is observed as 15 °C during winter and 45 °C in summer season.

2.2 Sensors Used in the Study

All those parameters are measured using three sets of sensors. Measurement of SO_2
has been done using SPEC DGS-SO2 and SPEC 3SP_SO2-20P which are electrochemical sensors, while the third sensor Horiba APSA 370 is based on ultraviolet
fluorescent (UVF) method. Measurement of NO_2 has been done using Alphasense
NO2-B43F, SPEC 3SP_NO2_5FP (both electrochemical sensors), and Horiba
APNA 370 (chemiluminescence approach based) sensors. Particulate matter concentration has been monitored using Plantower PMS7003, Prana Air PAS-OUT-01 (both
based on light scattering principle), and Met One Instruments BAM 1020 (based on
beta attenuation method) sensors. Location-wise sensor setup is shown in Table 1.

2.3 Methodology

The methodology used in the current study has been depicted in Fig. 2. All the three
monitoring stations have been set up at the roof top of administrative block inside
the campus. The monitoring stations have been installed at an elevation of 312.45 m

Table 1 Sensors used in the study

Pollutant	Monitoring stations		
	MS-1	MS-2	MS-3
Particulate matter (PM$_{2.5}$, PM$_{10}$)	Plantower PMS7003	Prana air PAS-OUT-01	Met one instruments BAM 1020
Nitrogen dioxide (NO$_2$)	Alphasense NO2-B43F sensor	SPEC 3SP_NO2_5FP	Horiba APNA 370
Sulphur dioxide (SO$_2$)	SPEC DGS-SO2	SPEC 3SP_SO2, 20P	Horiba APSA 370

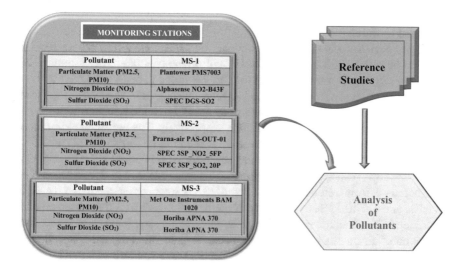

Fig. 2 Methodology

above MSL. Sensors measuring SO$_2$ and NO$_2$ at MS-1 have been started 30 min earlier than other two stations MS-2 and MS-3 as these sensors require preheating (warming up) before measurement based on their sensor characteristics. Other two stations have been turned on at the same time after 30 min of starting of MS-1. The MS-1 has been set up to monitor at 1-min intervals. MS-2 has been set up to monitor at 3-sec interval. MS-3 has been set up to monitor at 1-h interval. Results of the current study have been analyzed and compared with the previous studies [22, 25].

3 Results and Discussion

Data has been collected from 10 November 2021 to 20 November 2021 (10 days). The average value of PM$_{2.5}$ concentration obtained from MS-1 is found to be 31.38 µg/m^3. The average value observed from MS-2 and MS-3 is found to be 32.56 µg/m^3 and 28.59 µg/m^3. The average value of PM$_{10}$ concentration observed from MS-1, MS-2, and MS-3 is found to be 63.14 µg/m³, 55.69 µg/m³, and 61.9 µg/m³, respectively. Similarly, the average value of concentration of NO_2 obtained from MS-1, MS-2, and MS-3 is 14.60 µg/m³, 13.04 µg/m³, and 14.67 µg/m³. While the average SO_2 concentration obtained from MS-1, MS-2, and MS-3 is 12.70 µg/m³, 14.67 µg/m³, and 9.94 µg/m³, respectively. The results obtained in the current study have been compared with the previous studies of Deepak and Jaya [6] and Verma et al. [26]. Verma et al. [26] have made yearly analysis (2000–2014) of SPM, RSPM, NO_2, and SO_2 at industrial area. The concentration of NO_2 and SO_2 has been found to be below the standard limits. Deepak and Jaya [6] have measured yearly average concentration of $PM_{2.5}$ and PM_{10} for year 2016–2017 at residential area and traffic intersection. The residential area had higher concentration of $PM_{2.5}$ and PM_{10} when compared to traffic intersection area. The study of only particulate matters (PM_{10}, $PM_{2.5}$) has been done in the residential area. The percentage difference for $PM_{2.5}$ is found to be (−) 107.28% between MS-1 and reference study. The difference for MS-2 and MS-3 with reference study for $PM_{2.5}$ is found to be (−) 104.62% and (−) 113.74%, respectively. The difference between concentration values obtained from MS-1, MS-2, and MS-3 with reference study for PM_{10} has been observed to be (−) 124.18%, (−) 131.60%, and (−) 125.39%, respectively. The percentage difference for NO_2 between MS-1 and reference study is found to be (−) 55.90%. The difference for MS-2 and MS-3 with reference study for NO_2 is found to be (−) 66.39% and (−) 58.16%, respectively. The difference between MS-1, MS-2, and MS-3 with data obtained from reference study for SO_2 has been observed to be 14.81%, 28.59%, and (−) 10.12%, respectively. Negative value indicates that observed concentration is less than reference study data. The main reason for obtaining less concentration of air quality parameters from sensors is location of sensor installation. Due to the location, the sensors are less susceptible to the pollution caused by traffic and other anthropogenic activities. The main reason for conducting current study for short period is difficulty faced during setting up and calibration of sensors of MS-1 and MS-2 (Table 2).

The minimum value of $PM_{2.5}$ concentration obtained from MS-1 is 25.78 µg/m³, while the maximum value is found to be 37.01 µg/m³. For MS-2, the minimum value of $PM_{2.5}$ concentration is 27.19 µg/m³ and the maximum value is 36.69 µg/m³. Similarly for MS-3, the minimum and maximum ranges are 23.51 and 33.63 µg/m³, respectively. The variation in the obtained values may be due to less vehicular movement and other anthropogenic activities near the monitoring stations. Figure 3 shows the variation of $PM_{2.5}$ concentration obtained from MS-1, MS-2, and MS-3.

The minimum and maximum range of PM_{10} concentration obtained from MS-1 is 43.21 µg/m³ and 74 µg/m³. For MS-2, the minimum value of PM_{10} concentration is 38.52 µg/m³ and the maximum value is 66.90 µg/m³, while for MS-3, the minimum and maximum ranges are 41.52 and 72.61 µg/m³, respectively. The variation in the obtained values may be due to less vehicular movement and other anthropogenic activities near the monitoring stations. Figure 4 shows the variation of PM_{10} concentration obtained from MS-1, MS-2, and MS-3.

Table 2 Average concentration of pollutants

Parameter	MS-1 concentration (24 h. average) (μg/m^3)	MS-2 concentration (24 h. average) (μg/m^3)	MS-3 concentration (24 h. average) (μg/m^3)	Concentration obtained from reference study (μg/m^3)				
				Deshmukh et al. [8] Residential area	Verma et al. [25] Industrial area	Deepak and Jaya [6] Residential area	Deepak and Jaya [6] Traffic intersection	Sinha [22] Industrial area
PM$_{2.5}$	31.38	32.56	28.59	225.1	–	104	83	–
PM$_{10}$	63.14	55.69	61.9	368.2	–	270	190	227.50
NO$_2$	14.64	13.04	14.16	–	26	–	–	37.81
SO$_2$	12.76	14.67	9.94	–	11	–	–	14.41

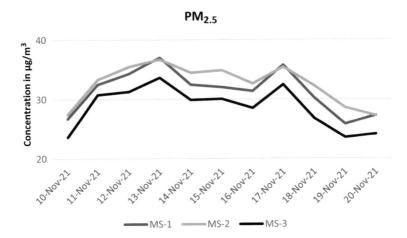

Fig. 3 PM$_{2.5}$ concentration

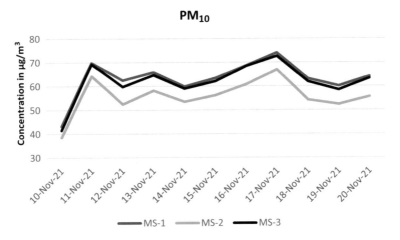

Fig. 4 PM$_{10}$ concentration

The minimum value of NO$_2$ concentration obtained from MS-1 is 13.25 μg/m^3, while the maximum value is found to be 15.74 μg/m^3. For MS-2, the minimum value of NO$_2$ concentration is 11.37 μg/m^3 and the maximum value is 15.19 μg/m^3. Similarly for MS-3, the minimum and maximum ranges are 11.52 and 15.39 μg/m^3, respectively. The variation in the obtained values may be due to less vehicular movement near the monitoring stations. Figure 5 shows the variation of NO$_2$ concentration obtained from MS-1, MS-2, and MS-3.

The minimum and maximum range of SO$_2$ concentration obtained from MS-1 is 9.82 μg/m^3 and 16.86 μg/m^3. For MS-2, the minimum value of SO$_2$ concentration is 10.92 μg/m^3 and the maximum value is 19.53 μg/m^3, while for MS-3, the minimum and maximum ranges are 13.78 and 8.74 μg/m^3, respectively. The variation in the

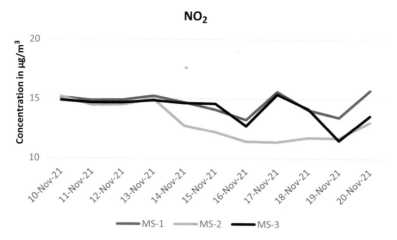

Fig. 5 NO₂ concentration

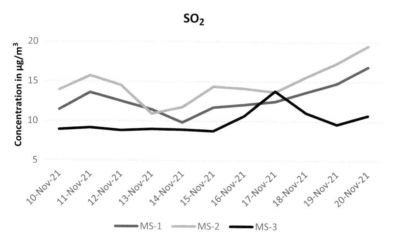

Fig. 6 SO₂ concentration

obtained values may be due to fuel combustion, and also there is no power plant industry near the monitoring stations. Figure 6 shows the variation of SO_2 concentration obtained from MS-1, MS-2, and MS-3.

4 Conclusions

Comparison of air pollutant (PM_{10}, $PM_{2.5}$, NO_2, and SO_2) measurement obtained from three monitoring stations MS-1, MS-2, and MS-3 has been carried out at same elevation at NIT Raipur. Continuous data of 10 days (10 November 2021 to 20

November 2021) has been recorded and considered for comparison. The results obtained in the current study have been compared with the previous studies. The percentage difference for $PM_{2.5}$ is found to be $(-)$ 107.28% between MS-1 and reference study. The difference for MS-2 and MS-3 with reference study for $PM_{2.5}$ is found to be $(-)$ 104.62% and $(-)$ 113.74%, respectively. The difference between concentration values obtained from MS-1, MS-2, and MS-3 with reference study for PM_{10} has been observed to be $(-)$ 124.18%, $(-)$ 131.60%, and $(-)$ 125.39%, respectively. The percentage difference for NO_2 between MS-1 and reference study is found to be $(-)$ 55.90%. The difference for MS-2 and MS-3 with reference study for NO_2 is found to be $(-)$ 66.39% and $(-)$ 58.16%, respectively. The difference between MS-1, MS-2, and MS-3 with data obtained from reference study for SO_2 has been observed to be 14.81%, 28.59%, and $(-)$ 10.12%, respectively. From the results obtained after comparing with previous studies, it can be concluded that the gas sensors can also be used to carry out the concentration measurement study as it gives satisfactory results.

References

1. Bing, L., Qing-Hao, M., Jia-Ying, W., Biao, S., Ying, W.: Three-dimensional gas distribution mapping with a micro-drone, pp. 6011–6015. 2015 34th Chinese Control Conference (CCC) (2015)
2. Borrego, C., Costa, A.M., Ginja, J., Amorim, M., Coutinho, M., Karatzas, K., Sioumis, T., Katsifarakis, N., Konstantinidis, K., De Vito, S., Esposito, E.: Assessment of air quality microsensors versus reference methods: The EuNetAir joint exercise. Atmos. Environ. **147**, 246–263 (2016)
3. Borrego, C., Ginja, J., Coutinho, M., Ribeiro, C., Karatzas, K., Sioumis, T., Katsifarakis, N., Konstantinidis, K., De Vito, S., Esposito, E., Salvato, M.: Assessment of air quality microsensors versus reference methods: The EuNetAir Joint Exercise – Part II. Atmos Environ. **193**, 127–142 (2018)
4. Carrozzo, M., Vito, S.D., Esposito, E., Formisano, F., Salvato, M., Massera, E., Francia, G.D., Delli Veneri, P., Iadaresta, M., Mennella, A.: An UAV mounted intelligent monitoring system for impromptu air quality assessments. In: Convegno Nazionale Sensori, pp. 497–506. Springer, Cham (2018)
5. Castell, N., Dauge, F.R., Schneider, P., Vogt, M., Lerner, U., Fishbain, B., Broday, D., Bartonova, A.: Can commercial low-cost sensor platforms contribute to air quality monitoring and exposure estimates? Environ. Int. **99**, 293–302 (2017)
6. Deepak, S., Jaya, D.: Seasonal variations in mass concentrations of PM10 and PM2. 5 at traffic intersection and residential sites in Raipur city. Res. J. Chem. Environ. **22**, 25–31 (2018)
7. Desai, N.S., Alex, J.S.R.: IoT based air pollution monitoring and predictor system on Beagle bone black. In: 2017 International Conference on Nextgen Electronic Technologies: Silicon to Software (ICNETS2), pp. 367–370. IEEE (2017)
8. Deshmukh, D.K., Deb, M.K., Mkoma, S.L.: Size distribution and seasonal variation of size-segregated particulate matter in the ambient air of Raipur city, India. Air Qual. Atmos. Health. **6**(1), 259–276 (2013)
9. Devarakonda, S., Sevusu, P., Liu, H., Liu, R., Iftode, L., Nath, B.: Real-time air quality monitoring through mobile sensing in metropolitan areas. In: Proceedings of the 2nd ACM SIGKDD international workshop on urban computing, pp. 1–8 (2013)

10. Ferreira, F., Tente, H., Torres, P., Cardoso, S., Palma-Oliveira, J.M.: Air quality monitoring and management in Lisbon. Environ. Monit. Assess. **65**(1), 443–450 (2000)
11. Hasenfratz, D., Saukh, O., Walser, C., Hueglin, C., Fierz, M., Arn, T., Beutel, J., Thiele, L.: Deriving high-resolution urban air pollution maps using mobile sensor nodes. Pervasive Mob Comput. **16**, 268–285 (2015)
12. Hu, Z., Bai, Z., Yang, Y., Zheng, Z., Bian, K., Song, L.: UAV aided aerial-ground IoT for air quality sensing in smart city: Architecture, technologies, and implementation. IEEE Network. **33**(2), 14–22 (2019)
13. Kadri, A., Yaacoub, E., Mushtaha, M., Abu-Dayya, A.: Wireless sensor network for real-time air pollution monitoring. In: 2013 1st international conference on communications, signal processing, and their applications (ICCSPA), pp. 1–5. IEEE (2013)
14. Kumar, P., Morawska, L., Martani, C., Biskos, G., Neophytou, M., Di Sabatino, S., Bell, M., Norford, L., Britter, R.: The rise of low-cost sensing for managing air pollution in cities. Environ. Int. **75**, 199–205 (2015)
15. Ma, Y., Richards, M., Ghanem, M., Guo, Y., Hassard, J.: Air pollution monitoring and mining based on sensor grid in London. Sensors. **8**(6), 3601–3623 (2008)
16. Mead, M.I., Popoola, O.A.M., Stewart, G.B., Landshoff, P., Calleja, M., Hayes, M., Baldovi, J.J., McLeod, M.W., Hodgson, T.F., Dicks, J., Lewis, A.: The use of electrochemical sensors for monitoring urban air quality in low-cost, high-density networks. Atmos. Environ. **70**, 186–203 (2013)
17. Penza, M., Suriano, D., Pfister, V., Prato, M., Cassano, G.: Urban air quality monitoring with networked low-cost sensor-systems. Proc. (MDPI). **1**(4), 573 (2017)
18. Peters, J., Van den Bossche, J., Reggente, M., Van Poppel, M., De Baets, B., Theunis, J.: Cyclist exposure to UFP and BC on urban routes in Antwerp, Belgium. Atmos. Environ. **92**, 31–43 (2014)
19. Postolache, O.A., Pereira, J.D., Girao, P.S.: Smart sensors network for air quality monitoring applications. IEEE Trans. Instrum. Meas. **58**(9), 3253–3262 (2009)
20. Riza, N.A., Sheikh, M.: All-silicon carbide hybrid wireless-wired optics temperature sensor network basic design engineering for power plant gas turbines. Int J Optomechatronic. **4**(1), 83–91 (2010)
21. Rohi, G., Ofualagba, G.: Autonomous monitoring, analysis, and countering of air pollution using environmental drones. Heliyon. **6**(1), e03252 (2020)
22. Sinha, D.: Ambient air quality status in an industrial area of Raipur city in the year 2015. J. Appl. Chem. **7**(3), 649–655 (2018)
23. Snyder, E.G., Watkins, T.H., Solomon, P.A., Thoma, E.D., Williams, R.W., Hagler, G.S., Shelow, D., Hindin, D.A., Kilaru, V.J., Preuss, P.W.: The changing paradigm of air pollution monitoring. Environ. Sci. Technol. **47**(20), 11369–11377 (2013)
24. Thu, M.Y., Htun, W., Aung, Y.L., Shwe, P.E.E., Tun, N.M.: Smart air quality monitoring system with LoRaWAN. In: 2018 IEEE International Conference on Internet of Things and Intelligence System (IOTAIS), pp. 10–15. IEEE (2018)
25. Verma, M.K., Patel, A., Sahariah, B.P., Choudhari, J.K.: Computation of air quality index for major cities of Chhattisgarh state. Environ. Claim. J. **28**(3), 195–205 (2016)
26. Wen, T.H., Jiang, J.A., Sun, C.H., Juang, J.Y., Lin, T.S.: Monitoring street-level spatial-temporal variations of carbon monoxide in urban settings using a wireless sensor network (WSN) framework. Int. J. Environ. Res. Public Health. **10**(12), 6380–6396 (2013)

Encoding Social Media Wording Indexes to Analyze PM$_{2.5}$ Problem Perception

Sureeporn Nipithwittaya

Abstract The purpose of this research was to create a database of areas with PM$_{2.5}$ and to map the spatial and temporal PM$_{2.5}$ density from a social media wording index. In addition, the study intends to determine the perception of PM$_{2.5}$ through text from social media so that people can find ways to protect themselves from the impact of PM$_{2.5}$. The information acquired is from image data and location data in eight provinces of the upper northern region of Thailand—namely, Mae Hong Son, Chiang Mai, Chiang Rai, Phayao, Nan, Lampang, Phrae and Lamphun. According to the search, the terms define the repetition-frequency index for data analysis. It can be used to analyze and map the PM$_{2.5}$ density and can show the spatial PM$_{2.5}$ in the study area. Moreover, the results of the study reflect that people are alert and aware of PM$_{2.5}$ problems through social media text. This includes the views and opinions of the people about the designated sources and effects of PM$_{2.5}$ that the first comes from open burning. It is comprised of the impact that occurs with both the visibility of the vision's impact on quality of life, health and environment. People are aware of the problems and the consequences that have arisen and have experienced such problems. Guidelines for self-solution of problems are things such as using a PM$_{2.5}$ mask and going into the open air during the PM$_{2.5}$ crisis. Moreover, Chiang Mai and Chiang Rai provinces are where individuals suffer from PM$_{2.5}$ problems and are the most discussed on social media, respectively. This corresponds to air-quality standard reports from the local stations.

Keywords PM$_{2.5}$ · Word index · Social media · Density map · Air pollution

S. Nipithwittaya (✉)
Srinakharinwirot University, Bangkok, Thailand
e-mail: Sureepornn@g.swu.ac.th

101

1 Introduction

The situation of $PM_{2.5}$ is worse than the levels of the standard. This is a serious environmental crisis and a major problem affecting health and quality of life. The main cause of $PM_{2.5}$ generation is from human activities. Two main sources of $PM_{2.5}$ urban emissions are: industrial production and transportation [1], At the time of this study, the main emissions in rural areas were caused by burning garbage, wildfires and slash and burn. Haze and air pollution are important problems in several countries, especially the Northern part of Thailand [2] that has eight provinces—namely, Mae Hong Son, Chiang Mai, Chiang Rai, Phayao, Nan, Lampang, Phrae and Lamphun; there people are suffering from one of the worst levels from $PM_{2.5}$ for a long time, which occurs widely and annually [3]. This problem causes poor visibility, affecting land traffic and lodging. It also affects health, causing problems for the respiratory system and the daily livelihood of the people [2]. Therefore, reducing $PM_{2.5}$ is the key task for improvement of air quality [1].

Currently, social media is very popular because everyone can be both a receiver and a sender. It can be a very useful tool that plays a role in quick and timely communication. In addition, it is a channel for exchange of knowledge and ideas of the new generation who join together as a group or socially through a website such as camping lovers and/or dog lovers. Additionally, index-based wording has spatial geotag data throughout social media; it comes from the fact that users share image information or comment on various commentaries related to a place that can be used to analyze density and to describe spatial distribution [4]. Word indexes with great similarities in each group have high sentimental values that can be used in planning, solving problems and development in various areas.

Social media can be used to monitor and help people to be aware of the effects of $PM_{2.5}$, including that the public can access a situation in real time; thus, innovation and technology can be used to manage the environmental quality and also enable people to participate in informing the government to solve problems in a timely manner. Moreover, recognizing information enables people to prepare and protect themselves from the effects of $PM_{2.5}$. Then, this study has three objectives: (1) to create a database of areas with $PM_{2.5}$ from the social media wording index, (2) to map the density of spatial and time $PM_{2.5}$, and (3) to study the perception of $PM_{2.5}$ through comments from social media.

2 Methodology and Data

Studying and analyzing was done by using social media data through Facebook and Twitter applications based on users sharing information, images or comments related to research areas during 2019–2020. Research planning and field collection was carried out in eight provinces of the Northern of Thailand. This research is divided into two main processes. First, data and tools were collected by searching for appropriate data in both the content and the volume it used in the study; and second, analyzing and screening data for spatial and time relationships to make a density map and information was classified as follows: primary data includes PM$_{2.5}$ data from criteria set through the social media Facebook application. Secondary data, research and data collection of PM$_{2.5}$ was obtained from research papers and related articles.

Tools used to store, collect and analyze data are as follows: operating programs, including Microsoft Excel and Quantum GIS. For data collection, social media indexes were searched for detailed information: the researchers gathered data by probing social media, Facebook and Twitter to search for public information from uploads of PM$_{2.5}$ by data users, both image and location information, which was subject to certain conditions. Queries were obtained from studies and compilations of repeated summaries. Designing and preparing databases in the design, and preparation of the preliminary database, included details from the extracted text that were classified as follows: name, place, coordinates (latitude, longitude) date, time and text.

Data extracted from Facebook was location information in conjunction with other information such as date, time, and pictures. But social media, Facebook and Twitter information was used to interact with others, as well as to choose to share information, images or videos publicly. The filtering process was divided into two categories: (1) filtered text that indicates the date, time, and level of impact of PM$_{2.5}$; and (2) filtered from users who had itemized it from social media information and text with misspellings or misrepresented place names and spatial information— namely, where it is mentioned, referenced from a location on Google Earth. Collected data on social media with PM$_{2.5}$ data was from Facebook and Twitter with surveys. Social media databases were collected from social media to create a file and save it to Microsoft Excel. Geotag databases were made of a Perception of PM$_{2.5}$ problem map and density of PM$_{2.5}$ map. The PM$_{2.5}$ maps were validated in all 8 study provinces. Quantitative text was analyzed with cloud-word to count-word frequency indexes. Comments and opinions were used for content analysis and included a summary and discussion of the results. The theoretical framework of this study follows in Fig. 1.

3   Results

3.1   PM$_{2.5}$ Database Based on Spatial Media Wording Index

Based on Table 1, the database was extracted and designed to identify location name, district, province and coordinates using two types of filtering processes as described in Sect. 2. A spatial database table was designed and created, as shown in Table 2. This table contains a spatial media wording index, text (comments and opinions), time, date, month and year.

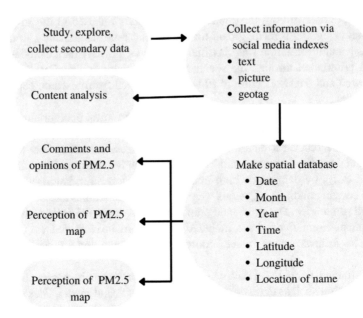

Fig. 1 The theoretical framework of this study

Table 1 Examples of PM$_{2.5}$ database from spatial media wording index

Location name	District	Province	X	y
Faculty of Business Administration, Chiang Mai University	Muang	Chiang Mai	18.794068	98.956777
Doi Suthep	Muang	Chiang Mai	18.803081	98.919840
Doi Luang Chiang Dao	Chiang Dao	Chiang Mai	19.397569	98.890016
Mae fah Luang university	Muang	Chiang Rai	20.044626	99.894075
Doi Nang non	Mae sai	Chiang rai	20.371706	99.852661
Mae Korn intersection	Muang	Chiang Rai	19.876750	99.831141
Baan Mae kopi	Khun Yuam	Mae Hong Son	18.751978	98.069128
Mae Sariang hospital	Mae Sariang	Mae Hong Son	18.162479	97.939932
Mortgage reservoir	Phu sang	Phayao	19.658366	100.354164
Phayao University	Muang	Phayao	19.030247	99.898065
Kwan Phayao	Muang	Phayao	19.163180	99.857097

Table 2 Example of PM$_{2.5}$ database from preliminary media wording index

Wording index	Text	Time	Date	Month	Year
Northern dust	Please give me some stimulation. It is a real crisis#ฝุ่นPM25 #ฝุ่นภาคเหนือ #ฝุ่นเชียงใหม่	13.30	28	March	2019
Northern dust	#แม่สาย #เชียงราย Can't the weather? Can't you stop burning? I can't do this anymore. #วิกฤติฝุ่นเชียงใหม่ #ฝุ่นเชียงราย #ฝุ่นภาคเหนือ #ฝุ่นPM25	12.50	30	March	2019
Chiang Rai dust	Beautiful view covered in dust, #ฝุ่นเชียงราย #ฝุ่นPM25	18.56	28	March	2019
Wildfire	Images straight from Doi on Chiang Mai can be burned. It is burned. It is hard to burn#เชียงใหม่ #pm25 #ไฟป่า #เผาป่า #chiangmai #รีวิวเชียงใหม่	20.09	19	March	2019
Smog	When the app measures dust; however, the words 'hideous' and 'unhealthy' Thai tourists just turn their heads away from Chiang Mai to get some air elsewhere. But Chiang Mai people can only open the window to see if they will 'see' Doi Suthep today. "It is a bioweapon," #เชียงใหม่. #หมอกควัน	10.52	30	March	2019

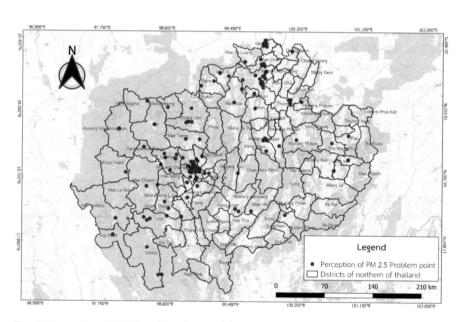

Fig. 2 Perception of PM2.5 problem from social media index in Northern Thailand

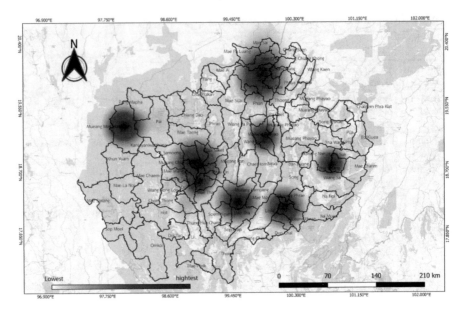

Fig. 3 Density of PM$_{2.5}$ social media index in Northern Thailand

3.2   Perceptions of PM$_{2.5}$ Problem from Social Media Index and the Density of PM$_{2.5}$ Mapping

Spatial and time PM$_{2.5}$ mapping were designed and made into spatial databases. Namely, the positions mentioned were referenced from locations on Google Earth to import geographic information systems. Coordinates were mapped perceptions of the PM$_{2.5}$ problem from the social media index and the density of PM$_{2.5}$ maps, as shown in Figs. 2 and 3. The results of the study in the Northern regions of eight provinces and the separate provinces are described in the following paragraphs.

The Pollution Control Department (PCD) has 15 air-quality monitoring stations in the eight Northern provinces in Thailand. Studies have shown that the public sector has a frequency of problem perception and the most mentioned PM$_{2.5}$ in the northern regions are in the Mueang District, especially in the Mueang Chiang Mai, Mueang Chiang Rai and Mae Sai District stations, respectively.

From Fig. 2, the study showed where people identified PM$_{2.5}$ problems, which, compared to the number of PCD air-quality monitoring stations, had one air-quality monitoring station in Lamphun Province at 68 t Station, Tambon Ban Klang, Amphoe Mueang, Lamphun Province. It was found that the positions indicated were greater than the number of measuring stations, and the perception of the problem was mainly in urban areas. Likewise, in Lampang Province, the PCD has 4 air-quality monitoring stations in Lampang Province—namely, 37 t Station, PhraBat Subdistrict, Mueang at 38 t, SopPad Subdistrict, Mae Mau at 39 t District, Ban Dong Subdistrict, Mae Mau District at 40 t, Mae Mau Subdistrict, Mae Mau District, Lampang Province. Public awareness was in accordance with the location where the air quality of the relevant authorities was measured.

Although in Mae Hong Son Province, the location of PM$_{2.5}$ was based on the transport route and was distributed in almost all districts of the province. Meanwhile, the PCD has one air-quality monitoring station in Mae Hong Son Province—namely, at 58 t Station, Jungkham Subdistrict, Mueang District and Mae Hong Son Province. The PCD has one air-quality monitoring station in Phrae Province—namely, at 69 t Station, Najak Subdistrict, Mueang District, Phrae Province. Accordingly, the social media locations are consistent with the locations of air-quality monitoring stations. Although in Chiang Mai, the public sector has identified PM$_{2.5}$ problems in almost all districts and they are very concentrated, especially in Mueang District, Chiang Mai Province, Mueang 030 Subdistrict, Mueang 031, Thep Rattana Hospital, Mae Chaem District, O33 Subdistrict, Mae Na Subdistrict, Chiang Dao District in Phayao Province. The PCD has one air-quality monitoring station in Phayao Province at 70 t Station, Tambon Ban Tom, Amphoe Mueang, Phayao Province.

Analysis of locator data showed that the Kwan Phayao area, where there is an impact of PM$_{2.5}$, was indicated in the province. In Chiang Rai, the impact of PM$_{2.5}$ is indicated in the provinces in Mueang district and Mae Sai district, which corresponds to the monitoring station where the PCD has two air-quality monitoring stations: at 57 t Station, Wiang Subdistrict, Mueang at 73 t District, Wiang Pang Kham Subdistrict, Mae Sai District, Chiang Rai Province. Nan showed an impact of PM$_{2.5}$ that was identified in 4 districts, whereas the PCD has two air-quality monitoring stations in Nan Province at 67 t Station, Nai Wiang Subdistrict, Mueang at 75 t, Huai Khon Subdistrict, Chaloem Phra Kiat District, Nan Province.

The density of the locator from the social media query index covers all eight provinces and density locations are formed in urban areas, which are characterized by flat terrain, valleys and basins. Such terrain is another important factor that contributes to the accumulation of PM$_{2.5}$, especially during the winter to dry season [5], mainly in the Chiang Mai basin.

3.3　Studying the Perception of PM$_{2.5}$ Through Social Media

Studying of the perception of PM$_{2.5}$ through social media comments is important so that people can find ways to protect themselves from its effects. Text analysis uses content analysis to systematically summarize and distinguish the characteristics of text. The results of the data analysis show the details in the following sections.

3.3.1　The PM$_{2.5}$ Problem Situation

One poster said:

"Encourage Chiang Mai to get the smoke pollution gone soon #pm25 #ขออากาศ บริสุทธิ์."

Another poster said:

"In the afternoon, the PM$_{2.5}$ bill hit almost six times higher than the morning, exceeding the benchmark and the highest of the dust values measured in the nine northern provinces, #ฝุ่นเชียงใหม่ #ThaiPBS."

3.3.2 Causes and Effects of PM$_{2.5}$

One poster said:

"Many times when I see villagers #เผาหญ้า #เผาป่า for agricultural hope, they can't stop cars and explain the problems that will follow, let's #รักษาป่า for our children."

Another poster said:

"Pictures from Doi on Chiang Mai, burned together, burned well, diligently burned, #เชียงใหม่ #pm25 #ไฟป่า #เผาป่า #chiangmai #รีวิวเชียงใหม่."

3.3.3 Problem Awareness

One of the posters said:

"No one cares about the lungs of the people in the north, burning forests and dust, not knowing what to burn, where the fire is going to go, where the dust is going to go viral, and where is the governor's province, where is the dust, there will be no dust for the sun, there will be no progress, please, pity the student lungs, #ฝุ่นละออง PM$_{2.5}$ #เผาป่า #ทวิตดีคนรีนัอย."

Another post said:

"Normal picture, no filter, sunny, barely seeing the sun, a lot of dust really #saveแม่สาย #ฝุ่นพิษ."

3.3.4 Solutions

Two posters said:

"In Doi Tao District, Chiang Mai, there has been rainfall for about an hour and a half, now there is thunder." "The information from Chiang Mai government office is a good sign, hopefully it will fall in the city and surrounding areas sometime #ฝนตก #ฝุ่นพิษ #เชียงใหม่."

Another poster said:

"The government should legislate to ban the consumption and trading of #เห็ดถอบ or #เห็ดเผาะ because villagers burn forests for mushrooms, for we consider it a problem that #ฝุ่นเชียงใหม่ #เผาป่า #pm25 #เชียงใหม่ is partly the root cause."

3.3.5 Self-Protection

One poster said:

"The smoke is back, all the photos taken today between 11:00 and 11:30, the images are raw files from the camera, it is recommended that everyone wear a mass that can keep out PM$_{2.5}$ all the time outside the building better (continued) #ฝุ่นละออง PM$_{2.5}$ #ฝุ่นเชียงราย #ฝุ่นภาคเหนือ #ฝุ่นเชียงใหม่."

Another poster said:

"Dense smog, worrying. Northern brothers, take care of yourself, #ภาคเหนือ #ไฟป่า #หมอกควัน."

4 Conclusion and Discussion

Air-quality monitoring stations have played a critical role in providing valuable datasets for recording air pollution [6]. But it takes a lot of budgets, so collecting of data by searching social media, Facebook and Twitter applications was used to pursue public information from users uploading PM$_{2.5}$ data. This included both images and geotagged data in the 8 provinces of Northern Thailand, including Mae Hong Son, Chiang Mai, Chiang Rai, Phayao, Nan, Lampang, Phrae and Lamphun. It can be analyzed and mapped for perceptions of PM$_{2.5}$ problems from a social media index and the density of PM$_{2.5}$ maps in the study area. In addition, the results reflect that the public is alert and aware of PM$_{2.5}$ problems through social media posts [7]. This is an effective tool for assessing social networking usage and validated sites plays role in environmental management.

The results of the study revealed the public's views and opinions on the determination and impact of PM$_{2.5}$ problems, which is consistent with research by Chairattanawan and Patthirasinsiri [8]. The results showed that the sources of PM$_{2.5}$ first were generated from outdoor burning consistent with this research; it has discussed the origin of PM$_{2.5}$ as from open-air burning, as well as the impact on visibility. Impact on quality of life and health and environment are important as well. Data also revealed that the public is aware of the problems and consequences, including suffering from them. Some people have taken steps to solve personal problems such as the use of PM$_{2.5}$ masks and not going out during a PM$_{2.5}$ dust crisis. According to most of the respondents [9], often they chose to wear regular or face cloth masks when going outdoors or to public gatherings.

In addition, the results of the study revealed that PM$_{2.5}$ is an annual problem in Northern Thailand and intensifies during the winter and the dry season [10]. Biomass burning emissions (i.e., agricultural and forest) are one of the major contributors to air pollution all over Thailand. Besides high pressure results in no air movement (wind), as well as the terrain of the North, with high mountain ranges interspersed with valleys, plains and urban areas such as Mueang Chiang Mai, Thailand. Here the basins are characterized by sinking air, allowing PM$_{2.5}$ to stay in the air for longer. Moreover population density is positively associated with PM$_{2.5}$ concentrations [1].

This study also revealed public opinion on the causes of PM$_{2.5}$ in Northern Thailand, which concluded two major causes: (1) wildfires in Thailand and neighboring countries; and (2) open-air burning, especially burning on farms or of weeds, of agricultural materials when preparing farmland. When analyzing spatial density based on data obtained from searches for the wording index, Chiang Mai is the province that suffers from $PM_{2.5}$ and is the most commonly discussed on social media. The next, Chiang Rai, which corresponds to the report of air-quality standards from local measuring stations, found that on March 16, 2019, it was reported that the $PM_{2.5}$ value there was at the highest level in the world, exceeding the standard level. Although Thailand's atmospheric air-quality standards currently are set to the revised standard, as announced in the *Gazette*; thus, setting an average of $PM_{2.5}$ for 24 hours, not exceeding 50 mg/m^3. This exceeds the World Health Organization (WHO) standard, which sets $PM_{2.5}$ values at 25 mg/m^3, and Thailand has set an average $PM_{2.5}$ value for 1 year, not exceeding 25 mg/m^3. Meanwhile, WHO defines an average $PM_{2.5}$ for 1 year, not exceeding 10 micrograms/cubic meter [7]. Therefore, the government should set and announce the standard and processes to exert control and solve the problem in the long term.

According to this study, the policies should be formulated before air pollution has a worse impact [11]. There are policy recommendations, as follows: First, the use of environmental economics measures to solve the problem of $PM_{2.5}$ management in the study area—namely, a pollution tax, which is a tax that is charged to those who emit pollutants into the environment. It is used in many countries as an air pollution tax and a water pollution tax. It should be charged based on the amount or type of pollution emitted into the environment. Second, promote the use of waste materials instead of sintering by encouraging the production of biodegradable natural materials by supporting green tax expense. Third, use subsidies as a tool by the government to encourage entrepreneurs to invest in projects that promote, or jointly address $PM_{2.5}$, low-interest loans and tax incentives. Four, determine the role of the public sector in the participation of local citizens in the field of surveillance and notifications, including the impact on themselves and their families. Finally, the use of environmental studies, knowledge, understanding of problems and impacts, including disaster practices and guidelines, for preventing $PM_{2.5}$.

Acknowledgement This research was funded by the Faculty of Social Science, Srinakharinwirot University, Thailand.

References

1. Han, S., Sun, B.: Impact of population density on $PM_{2.5}$ concentrations. A case study in Shanghai, China. Sustainability, 1–17 (2019)
2. Leelasittikul, K., Yuenyongchaiwat, K., Buranapuntalug, S., Pongpanit, K., Koonkumchoo, P.: Effect of haze and air pollution on cardio-respiratory systems in Northern Thailand. Thammasat Med. J. **18**(3), 339–348 (2018)
3. Anuttara, H., Yanasinee, S., Nittaya, P., Vivat, K.: Analysis of air quality impacts on human health using the geoinformatics application: Chiang Rai Province. Appl. Environ. Res. **39**(3), 25–32 (2017)
4. Sitthi, A.: Tags mining analysis using geotagged online social media data. J. Soc. Sci. Facul. Srinakharinwirot Univ. **21**(1), 304–319 (2018); Oanh, K. A study in urban air pollution improvement in Asia AIT. Research Project submitted to JICA-Research Institute, Asian Institute of Technology, October. (2017)

5. Sakunkoo, P., Thonglua, T., Sangkham, S., Jirapornkul, C., Limmongkon, Y., Daduang, S., Tessiri, T., Rayubkul, J., Thongtip, S., Maneenin, N., Pimonsree, S.: Human health risk assessment of PM$_{2.5}$—Bound heavy metal of anthropogenic sources in the Khon Kaen Province of Northeast Thailand. Heliyon. **8**, 1–7 (2022)
6. Su, L., Gao, C., Ren, X., Zhang, F., Cao, S., Zhang, S., Chen, T., Liu, M., Ni, B., Liu, M.: Understanding the spatial representativeness of air quality monitoring network and its application to PM$_{2.5}$ in the mainland China. Geosci. Front. **13**, 1–9 (2022)
7. Khan, N.A., Azhar, M., Rahman, M.N., Akhtar, M.J.: Scale development and validation for usage of social networking sites during COVOD-19. Technol. Soc. **70**, 1–10 (2022)
8. Chairattanawan, K., Patthirasinsiri, N.: Emission source impact and problem solving and management on PM$_{2.5}$ in the Northern part of Thailand. J. Assoc. Res. **25**, 432–446 (2020)
9. Supornpraditchai, T.: Demographic characteristics, individual self-protective behavior and health self-assessment among the people in areas with high PM$_{2.5}$ concentration in Bangkok. Acad. J. North Bangkok Univ. **10**(2), 1–18 (2021)
10. Bran, S.H., Macatangay, R., Surapipith, V., Chotamonsak, C., Chantara, S., Han, A., Li, J.: Surface PM$_{2.5}$ mass concentrations during the dry season over northern Thailand: Sensitivity to model aerosal chemical schemes and the effects on regional meteorology. Atmos. Res. **277**, 1–13 (2022)
11. Ji, H., Wang, J., Meng, B., Cao, Z., Yang, T., Zhi, G.: Research on adaption to air pollution in Chinese: evidence from social media-based health sensing. Environ. Res. **210**, 1–11 (2022)

Noise Mapping of Different Zones in an Urban Area During Deepawali Festival

Vishal Kumar, A. V. Ahirwar, and A. D. Prasad

Abstract Noise pollution is one of the unforeseen pollutions which are not taken into serious note in the environment. Its regular monitoring is very much essential in the present environment as it affects those who are exposed to it. In this study, monitoring and mapping of noise during the Deepawali festival is carried out in order to understand the impact of crackers contributing to the higher levels of noise. Sampling has been done in commercial, silence, and residential zones covering central part of Raipur city towards the western side of the urban area. Observations have been done on 3 November 2021 (day before Deepawali), 4 November 2021 (Deepawali day), and 5 November 2021 (day after Deepawali). Average noise levels observed on pre-eve are 74.11 dBA, 75.90 dBA, 71.10 dBA, and 70.70 dBA; on eve are 76.30 dBA, 74.30 dBA, 73.30 dBA, and 78.30 dBA; and on post-eve are 72.80 dBA, 72.70 dBA, 70.60 dBA, and 67.10 dBA, respectively. Based on the observations, noise maps for all 3 days have been prepared using inverse distance weighted (IDW) interpolation method in ArcGIS. Results of L_{eq}, L_{max}, and L_{min} are plotted using GIS tools. Comparison with the standard limits has also been carried out. It is observed that the noise levels on all 3 days in all zones are exceeding the World Health Organization (WHO) and Central Pollution Control Board (CPCB), India norms. These studies are useful for policy makers to make plans to mitigate the noise pollution in an urban region to provide healthy environment for all creatures.

Keywords Noise mapping · Deepawali · Raipur · GIS

V. Kumar (✉) · A. V. Ahirwar · A. D. Prasad
Civil Engineering Department, National Institute of Technology, Raipur, Chhattisgarh, India
e-mail: avahirwar.ce@nitrr.ac.in

© The Author(s), under exclusive license to Springer Nature
Switzerland AG 2023
W. Boonpook et al. (eds.), *Applied Geography and Geoinformatics for Sustainable Development*, Springer Geography,
https://doi.org/10.1007/978-3-031-16217-6_8

1   Introduction

Unwanted sound reaching to human ear and causing discomfort is called as noise [4]. Noise is generated in the environment by various sources mainly by the traffic in the developing and developed country worldwide [7, 10]. Cities that are populated and having higher number of vehicles register maximum environmental complaints regarding noise pollution [11]. Annoyance, irritations, and damages to the auditory system are few problems that are faced by people exposed to high levels of noise [1]. Not only the adults but children are also among the most vulnerable group as they are most exposed to noise in the present situations due to rapid growth in developing country [9]. During the festival eve there are some special factors which elevate the levels of noise on the specified day. Serial bursting of crackers on Deepawali eve leads to the generations of high levels of noise which affects the exposed populations [5, 8]. Crackers contain harmful chemicals which contribute to the air pollution along with high levels of noise leading to noise pollution [2]. India is a land of festival where crackers are burnt not only on Deepawali but also on various eves (marriage, New Year, birthday, etc.). However, other eves are celebrated partly, but Deepawali eve is carried out countrywide.

Visual presentation of the noise levels of different locations on a single platform for better understanding and the presence of noise in the environment is called noise mapping. Noise map can also call as a tool which shows visual information of the sound over a specified geographical area [3]. Mapping of noise also helps in urban planning. It can also be called as shifting from traditional monitoring to modern method of monitoring noise levels [6]. There is no spatial study on noise available for the Raipur city during the festival eve. The present work aims to study the difference in noise levels of selected zones of Raipur city during the Deepawali eve of 2021. Noise mapping during the eve is the first ever study carried out in the city.

2   Methodology of the Study

Study was carried out in two phases: (a) collecting of noise levels before eve, during eve, and post-eve and (b) comparison and preparations of noise map using ArcGIS software (version 10.3.1). Both phases of the study have been described through flowchart as shown in Fig. 3.

2.1   Study Area

Raipur capital city of Chhattisgarh was selected for the study. It has the highest population density as compared to other cities of the Chhattisgarh state. It is located at 22°33′ to 21°14′N latitude and 82°06′ to 81 ° 38′E longitudes. Three different

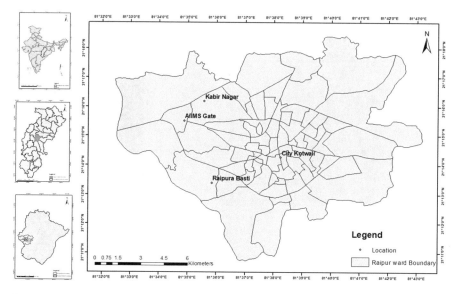

Fig. 1 Study area map during the Deepawali eve 2021

zones (residential, commercial, and silence) were selected for carrying out the study. A total of four locations were fixed among which two for residential (Kabir Nagar, Raipura), one for commercial (City Kotwali), and one for silence (AIIMS Gate, hospital). The locations of the stations have been shown by preparing the study area map using ArcGIS as shown in Fig. 1.

2.2 Methodology

Noise levels were collected using the sound level meter (Envirotech, SLM 100). Instrument used for taking the observations is shown in Fig. 2. The instrument was calibrated before taking readings, and then it was set up at various selected locations during the Deepawali eve. The latitudes and longitudes were also recorded which was used in preparation of the study area map. Observations were carried out at the four locations for 3 days, i.e., pre-eve, on eve, and post-eve. Noise levels were recorded for different times at different locations (City Kotwali 16:00–17:28 h, AIIMS Gate 18:00–19:16 h, Kabir Nagar 19:28–20:28 h, and Raipura 21:00–22:00 h) from 3 November to 5 November 2021. All the readings were taken by keeping the sound level meter 1.5 m above ground levels and following the noise rules 2000 (CPCB, New Delhi, India) protocols. The detailed methodology flowchart is shown in Fig. 3. After collections of data, average noise levels were found out as shown in Table 1. The maximum, minimum, and average L_{eq} values have been plotted for all the four locations during the eve. The plotted noise data is shown in Fig. 4. In ArcGIS 10.3.1 version, inverse distance weighted (IDW) interpolation method was used for

Fig. 2 Sound level meter
used in the study

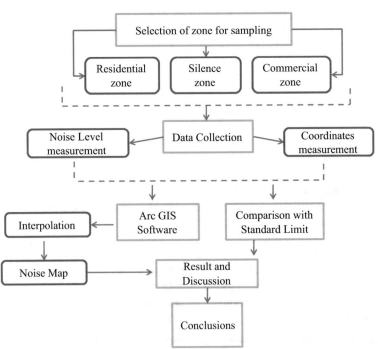

Fig. 3 Flowchart of methodology

Table 1 Observed noise levels during Deepawali eve 2021

Zone	Location and time	Average sound levels L_{eq}(dBA)			
		03/11/2021	04/11/2021	05/11/2021	CPCB limit
Commercial	City Kotwali (16:00–17:28)	74.10 (max 88.20) (min 61.10)	76.30 (max 90.40) (min 63.30)	72.80 (max 87.20) (min 68.00)	Daytime(65) Nighttime (55)
Silence	AIIMS gate (18:00–19:16)	75.90 (max 90.40) (min 60.60)	74.30 (max 83.90) (min 58.30)	72.70 (max 86.40) (min 57.70)	Daytime(50) Nighttime (40)
Residential	Kabir Nagar (19:28–20:28)	71.10 (max 88.30) (min 52.10)	73.30 (max 94.80) (min 53.70)	70.60 (max 86.30) (min 53.30)	Daytime(55) Nighttime (45)
Residential	Raipura (21:00–22:00)	70.70 (max 89.60) (miin 55.40)	78.80 (max 87.20) (min 54.60)	67.10 (max 76.30) (min 51.10)	Daytime(55) Nighttime (45)

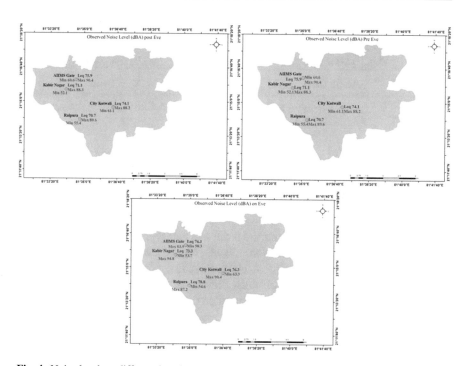

Fig. 4 Noise levels at different locations during eve

preparation of noise map for the city during the eve. The noise maps were prepared with the cell size 20 × 20 m. The prepared maps were exported, and comparison of noise levels for all the 3 days was done. The prepared noise maps have been shown in Fig. 6 for all the 3 days during the eve. The observed noise levels were also graphically compared with WHO and CPCB standard limits as shown in Fig. 5.

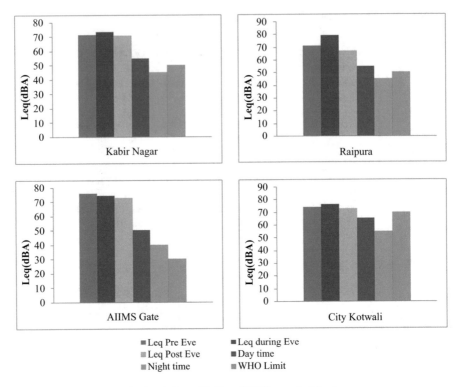

Fig. 5 Comparison of noise levels from CPCB and WHO standard limits

3   Results and Discussions

From the observations made for all the 3 days, it can be found out that during the eve of Deepawali, all the four locations showed higher levels of noise as compared to the pre- and post-eve. On third of November, i.e., pre-eve of Deepawali, City Kotwali recorded 74.10 dBA of noise level which is higher than post-eve (72.80 dBA). However it recorded higher value (76.30 dBA) than other 2 days during the eve as being a commercial zone few residences are also located in the area. Narrow road and dense population make this area prone to higher levels of noise. As the traffic was less on post-eve due to shutdown of shops during the festivals, it recorded lower noise level than pre-eve.

Consequently AIIMS Gate area recorded highest level (75.90 dBA) pre-eve due to the traffic as it is located nearer to national highway as well as small commercial area. Noise levels of 74.30 dBA were observed on eve. The main source of noise was crackers bursting and traffic. On post-eve this area also recorded lower noise level of 72.70 dBA. Major factor contributing to noise in this zone was traffic. As the traffic was low compared to pre-eve, lower noise was observed.

The two residential areas, i.e., Kabir Nagar and Raipura, recorded higher levels (73.30 dBA, 78.80 dBA) of noise on the Deepawali eve than pre- and post-eve,

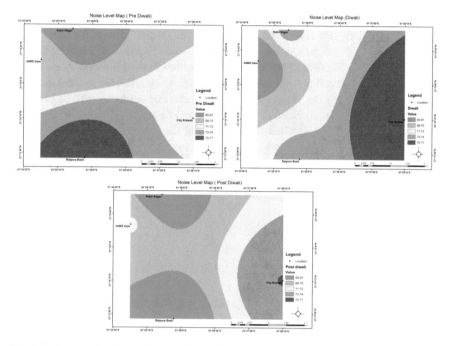

Fig. 6 Noise mapping during Deepawali eve 2021

respectively. The major source for noise in both the area was bursting of crackers. From the observation made on post-eve (70.60 dBA and 67.10 dBA) for both the region, it is found that lower noise was recorded.

From the above observations, it can be concluded that in residential areas higher levels were exceeded due to bursting of crackers and in other two areas (commercial and silence) maximum contribution to the higher noise levels was from the moving traffic as in post-eve both locations showed lower levels than pre-eve and festival eve. From the noise maps it is revealed that higher noise levels exist in all zones. On comparing the levels with the CPCB and WHO standard limits, it is revealed that all the locations showed higher levels of noise than the prescribed limit (commercial 65, 55; silence 50, 40; residential 55, 45 day- and nighttime in dBA) for all the zones. In Fig. 5 the comparison has been made to understand the difference in elevated noise levels. From the graphical comparison all the zones are at higher levels of noise for all the 3 days (Fig. 5) [12, 13].

4 Conclusion

The study done during the Deepawali eve for selected zones of Raipur reveals that there have been high levels on noise during festival due to burning of crackers. Maximum noise pollution was observed in the residential zones compared to the commercial and silence zones. The main source contributing to higher noise level in commercial and silence zone is traffic. Location of silence zone in the heart of the city is also one of the factors which makes the zone exposed to higher levels of noise. All the four locations showed higher noise levels than the WHO and CPCB prescribed limits. Hence there is a need for making strict bylaws to control the noise pollutions not only during the eve but also for other normal days as there has been high difference in the prescribed limit versus observed noise levels for all the zones. From the study it is revealed that all the zones of the city are prone to noise pollution. The prepared noise map will help in identifying the area with higher noise levels, and also it will help the policy makers for taking strict action in the identified region to mitigate the noise pollution. It is very much necessary to make our environment healthy so that the harmful effects of all the pollutants can be minimized on living beings.

References

1. Alam, W.: GIS based assessment of noise pollution in Guwahati City of Assam, India. Int. J. Environ. Sci. **2**(2), 2011 (2011)
2. Balashanmugam, P., Nehrukumar, V., Ramanathan, A., Balasubramanian, G.: A study on the influence of Deepavali festival on noise level in Chidambaram town, Tamilnadu, India. Res. J. Eng. Technol. **7**(1), 29 (2016)
3. de Noronha Castro Pinto, F.A., Moreno Mardones, M.D.: Noise mapping of densely populated neighborhoods—example of Copacabana, Rio de Janeiro—Brazil. Environ. Monit. Assess. **155**(1–4), 309–318 (2009). https://doi.org/10.1007/s10661-008-0437-9
4. Farooqi, Z.U.R., Sabir, M., Latif, J., Aslam, Z., Ahmad, H.R., Ahmad, I., Imran, M., Ilić, P.: Assessment of noise pollution and its effects on human health in industrial hub of Pakistan. Environ. Sci. Pollut. Res. **27**(3), 2819–2828 (2020). https://doi.org/10.1007/s11356-019-07105-7
5. Goswami, S., Swain, B.K., Mohapatra, H.P., Bal, K.K.: A preliminary assessment of noise level during Deepawali festival in Balasore, India. J. Environ. Biol. **36**, 981–984 (2013)
6. Kumar, R., Mukherjee, A., Singh, V.P.: Traffic noise mapping of Indian roads through smartphone user community participation. Environ. Monit. Assess. **189**(1), 17 (2017). https://doi.org/10.1007/s10661-016-5741-1
7. Lee, E.Y., Jerrett, M., Ross, Z., Coogan, P.F., Seto, E.Y.W.: Assessment of traffic-related noise in three cities in the United States. Environ. Res. **132**, 182–189 (2014). https://doi.org/10.1016/j.envres.2014.03.005
8. Patel, N.L., Bhave, P.P.: Study of noise pollution during Deepawali festival. Int. J. Innov. Res. Adv. Eng. **1**(6), 11 (2014)
9. Stansfeld, S., Clark, C.: Health effects of noise exposure in children. Curr. Environ. Health Rep. **2**(2), 171–178 (2015). https://doi.org/10.1007/s40572-015-0044-1
10. Tabraiz, S., Ahmad, S., Shehzadi, I., Asif, M.B.: Study of physio-psychological effects on traffic wardens due to traffic noise pollution; exposure-effect relation. J. Environ. Health Sci. Eng. **13**(1), 30 (2015). https://doi.org/10.1186/s40201-015-0187-x
11. Wu, J., Zou, C., He, S., Sun, X., Wang, X., Yan, Q.: Traffic noise exposure of high-rise residential buildings in urban area. Environ. Sci. Pollut. Res. **26**(9), 8502–8515 (2019). https://doi.org/10.1007/s11356-019-04640-1
12. W.H.O (World Health Organization). https://www.who.int/europe/health-topics/noise. Retrieved 07 Nov 2021
13. C.P.C.B (Central Pollution Control Board, New Delhi, India). https://cpcb.nic.in/noise-pollution/. Retrieved 07 Nov 2021

Digital Twins in Farming with the Implementation of Agricultural Technologies

Aakash Thapa and **Teerayut Horanont**

Abstract The increment in the human population along with the demand for food has caused a global food crisis and is expected to escalate in upcoming years. With the limited number of farmers, insufficient agricultural lands, pests in plants, climate change, and several other reasons that restrict food productivity, smart development of farming is necessary to overcome food scarcity. This chapter focuses on the importance of the implementation of digital twin farming platforms to sustain food security. We explore the involvement of artificial intelligence, the Internet of Things, big data, and cloud services to excel in farming using simulation, analysis, and accurate planning for growth in agricultural sectors. Our detailed review discusses the initial phase of digital twins in farming helping the experts or farmers to monitor agricultural scenes remotely to identify issues and come up with instant solutions for efficient food production. We also perform a case study on a digital twin paradigm in a solar energy-supplied farm and its contribution to two of the Sustainable Development Goals (SDGs): "Zero hunger" and "Affordable and clean energy." Furthermore, we outline the purpose of broad research in this domain for a sustainable future.

Keywords Digital twin · Internet of Things · Smart farming · Agriculture · Remote sensing

1 Introduction

The rising human population is expected to reach 9.3 billion by 2050 with the need to raise the production of food by 60% as stated by the Food and Agriculture Organization (FAO) [1]. Overcoming the food scarcity predicted in upcoming years is a global challenge and is also a part of the United Nation (UN)'s Sustainable Development Goals (SDGs). SDG Goal 2 which focuses on "Zero hunger" is one of

A. Thapa · T. Horanont (✉)
Sirindhorn International Institute of Technology, Thammasat University, Pathum Thani, Thailand
e-mail: m6322041135@g.siit.tu.ac.th; teerayut@siit.tu.ac.th

© The Author(s), under exclusive license to Springer Nature Switzerland AG 2023
W. Boonpook et al. (eds.), *Applied Geography and Geoinformatics for Sustainable Development*, Springer Geography,
https://doi.org/10.1007/978-3-031-16217-6_9

the aims of the UN to end hunger and sustain food security with improvement in nutrition [2]. Though world population is one of the reasons behind food scarcity, some other causes are climate change, weather, pests in plants, natural calamities, etc. [3, 4]. Thus, the integration of smart technologies has been introduced in recent years for effective food production in an efficient way.

Smart farming uses the technologies like Internet of Things (IoT) and cloud computing for gathering, processing, and analyzing the data to automate the system resulting in improvement in operation and farm management [5]. Digital farming, also known as smart farming, introduces big data analytics, IoT, information and communication technology, and the inclusion of decision-making artificial intelligence (AI) algorithms like deep learning and machine learning which has been a major advancement in the field of agriculture by qualitative and quantitative production with low cost [6].

The digital twin is a mirror representation of a physical object or a world in a digital form [7] and is a relatively new concept introduced in smart farming. Thus, our contribution is to explore the implementation of digital twins in the agricultural field from an initial phase and their importance for the advancement of smart farming to sustain food security. The novelty in this chapter comes from a case study on a zero-energy solar farm with the involvement of the digital twin model and a discussion on its contribution to SDGs.

The rest of the chapter is structured as follows: Sect. 2 provides the literature review of the digital twin in agriculture, which is substructured into Sect. 2.1 which provides the information on the development of the digital twin; Sect. 2.2 reviews about digital twin concept in agriculture from previous findings; Sect. 2.3 discusses simulation software and applications used in agriculture; Sect. 2.4 explains about SDGs achieved from digital twin. Section 3 studies on zero-energy digital twin paradigm and Sect. 4 concludes the chapter.

2   Digital Twin

Digital model, digital shadow, and digital twin appear as identical terms. However, there is a disparity in the level of data integration between physical and digital environments which is clarified in Fig. 1. Understanding in detail, the digital model does not have automated data flow in either way between the physical object, which is the real-world object, and the digital object. This signifies exchange in data is performed manually. In digital shadow, the automated flow of data from physical object to digital object is possible, but not vice versa. The changes in the physical environment can bring change to the digital environment. The digital twin has the capability that data flows automatically in both ways. The change in one state brings a change in another state. This means the digital object is also able to control the physical object's environment [8], which is important in smart farming. The integration of multiple technologies like big data, IoT, AI, etc. helps farmers or experts to perceive

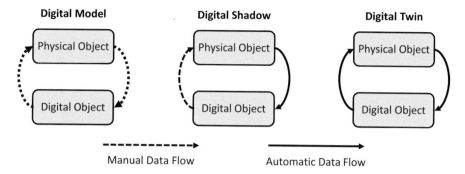

Fig. 1 Concept of the digital twin

the condition of the farms so they can adapt when there is a change in weather, climate, etc. [9].

2.1 Digital Twin in Smart Farming: Early Stage of Development

Digital twins in agriculture have recently caught a spike in attention. A search in Scopus using keywords "Digital Twin" *and* "Agriculture" returns just 63 results, most of which are published in the last 2 years. The current applications of the digital twin in farming are basic structures that need to take shape in the future. Since the digital twin is in the early stage of development, there have been very few studies in this domain. Some of the experimental studies are discussed below.

A recent study [9] develops a dashboard to display the information of data collected from soil probes. The probes are installed in the field to collect information on soil moisture, ground temperature, geospatial position, and air temperature and humidity, which are sent to an IoT agent to use in the cloud. From the gathered data, the amount of water is calculated to observe the water content. These cloud services produce dashboards for visualization after monitoring and analysis as shown in Fig. 2 in real time. The digital environment is created to make decisions and send the information back to the physical environment; however, it can only visualize the information to this date.

A feasibility study on a smart pig farm [10] introduces digital services able to control livestock farms as shown in Fig. 3. The sensors are used to compute temperature, ammonia content, carbon dioxide content, dust, humidity, etc. and sent to the big data component which decides to store data either in real time or non-real time. The big data, however, filters and stores the mandatory data needed for autonomous control analysis. The simulation component queries the data stored in the big data component. Then, the simulation is implemented and the outcome from the simulation suggests optimal carbon dioxide and temperature for the growth of

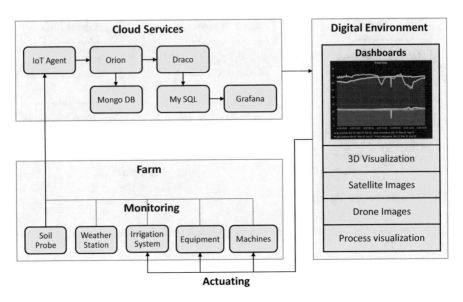

Fig. 2 System architecture of a DT for gathering data from the soil probe and dashboard to display the information [9]

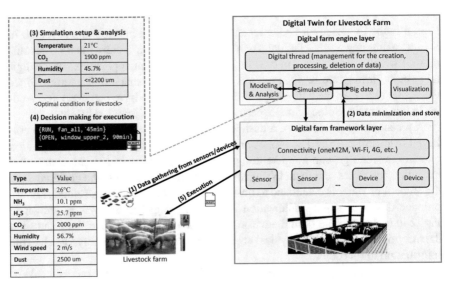

Fig. 3 Digital twin for a pig (livestock) farm [10]

livestock. The decision is made to control devices like fans and windows with the execution of script files written to control devices.

The detection of crop disease and nutrient deficiency was conducted [11] for better production with improved nutrients in crops to benefit consumers' health and more importantly food export. A low-cost but robust wireless sensor network called LoRaWAN is set up in farmland, and drone imagery is used. The cloud server is

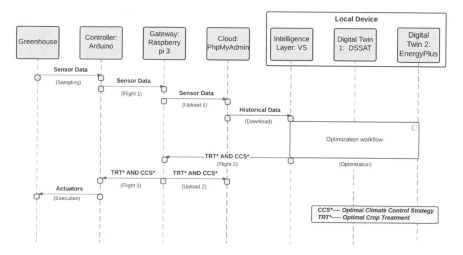

Fig. 4 A sequence diagram to explain communication from the greenhouse prototype to the digital twin to take measures for optimal climate control and crop treatment [12]

responsible for running machine learning algorithms to detect crop disease and nutrient deficiency automatically which can be remotely accessed by the farmers. The use of IoT technology helps in the constant monitoring of soil and results in automatic plant disease detection to recommend proper treatment for high production.

A test on a controlled environment, i.e., greenhouse prototype [12], is conducted to optimize productivity by controlling climate and crop treatment. The prototype is equipped with a greenhouse, fan, temperature and humidity sensor, irrigation system, controller, and radiation shield. The controller reads the inputs and the data is transmitted through the gateway to store in storage, where both the current and historical data are stored. The intelligence layer is responsible for choosing climate control strategies and crop treatment based on the knowledge. EnergyPlus[1] predicts the climate inside the greenhouse using external and internal temperature and humidity and allows control strategies optimization for controlling microclimate [13], whereas Decision Support System for Agrotechnology Transfer (DSSAT)[2] allows treatment optimization of crops [14] by providing required nutrients, water, etc. using historical and predicted microclimate data. Figure 4 is a sequence diagram for explaining the mechanism of communication between the devices for providing optimal climate control and crop treatment.

From the experiments conducted, we can acknowledge that IoT, cloud services, and AI algorithms are used in digital twin for collecting data, storing historical and current data, automatic analysis, visualization, and simulation. The observations indicate that monitoring the situation on the farm is possible; however, automated

---

[1] https://energyplus.net/

[2] https://dssat.net/

Table 1 Overview of some experiments in the digital twin for smart farming

Concept	Goal	Digital twin type	References
Water level monitoring	Observing water content using IoT technologies installed in soil	Monitoring, recollection	[9]
Livestock farming	Controlling devices to maintain optimal CO_2 and temperature	Monitoring, recollection, autonomous	[10]
Plant disease detection	Automatic plant disease and nutrient deficiency detection	Monitoring, recollection, prescriptive	[11]
Greenhouse	Microclimate control and crop treatment using simulation software	Monitoring, recollection, autonomous, predictive, prescriptive	[12]

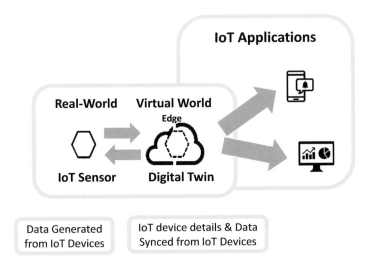

Fig. 5 Digital twin concept in IoT [16]

control in the farms with remote access is barely used and is in the phase of development. So digital twins can be identified in different approaches according to the objective of the research. The digital twins could be described as monitoring digital twin, imaginary digital twin, prescriptive digital twin, predictive digital twin, autonomous digital twin, or recollection digital twin [15]. Table 1 is the overview of the experiments we discussed earlier in this section, which also differentiate the type of digital twins based on the research objective.

2.2 Digital Twin Concept for Agriculture

The concept of the digital twin can be acknowledged in Fig. 5 in general, with the understanding from the above section. The real-world data is captured by IoT sensors and cameras, including drone images and satellite images, in the form of historical data and near real-time data. These data are used for monitoring, analyzing, and visualization. The AI part could be involved in the prediction of crop health and optimization. From the observations, the control of real-life objects from the virtual world could be done remotely with the help of software. However, the digital twin in the agricultural scenario is in the development phase. Nevertheless, there are already developed applications used in some of the research works. In Fig. 4, DSSAT and EnergyPlus are two software used to control the microclimate inside the greenhouse, which communicates with Python-based IDEs. Similarly, the script file can be executed to control livestock farms as shown in Fig. 3. We will discuss applications/software used for monitoring or controlling physical objects in the next section.

2.3 Simulation Software and Applications for Digital Farming

Simulation software and applications are used in digital environments to operate remotely for monitoring and controlling physical environments. EnergyPlus is an open-source simulation software that is used for operations like cooling, heating, lighting, ventilation, and water use. This program is capable of comparing more than one simulation result graphically [17]. Another software application, DSSAT is developed for crop growth simulation. This app is supported by 42 crops to this date and has different tools for soil, weather, genetics, and crop management [18]. The integration of EnergyPlus and DSSAT for production optimization can be seen in the digital twin implementation on a greenhouse prototype [12]. In a smart livestock farming feasibility study, it is described how digital twins can be used for visualizing information in 2D or 3D in a user-friendly interface. In addition, a script file can be executed for controlling the temperature and CO_2 level recommended by simulation results [10]. However, the test on smart livestock has not been fully conducted as it is in the phase of development. Figure 2 also shows that the progress is completed up to a dashboard in a digital environment. The digital twin is not fully developed because it is not possible to remotely act in a physical environment for water management [9]. A similar scenario is mirrored in [11] for plant disease and nutrient deficiency detection. The AI database can be able to provide information about plant diseases but is not able to act remotely. However, new technologies are introduced and expected to be introduced for remotely farming with precision.

FarmBot[3] is an open-source web application developed for configuring and controlling farms remotely. Figure 6 is a FarmBot application that controls a small farm based on the location of the plants. MQTT gateway is used to connect the application with the devices used on the farm. Raspberry Pi helps in downloading and running real-time or scheduled events on the farm and uploading sensor data from the farm. FarmBot digital twin is a 3D simulation of a physical garden. It has a built-in

---

[3] https://farm.bot/

Fig. 6 FarmBot web-app interface

camera for taking photos to create a 3D model using photogrammetry techniques [19]. FarmBot users can easily monitor and control the farm. This concept can take proper care of plants from different areas and contribute to food security [20].

These applications and software we discussed earlier are contributing to food security. They are also very important for controlling the digital twin model and are a part of digital twins. In conjunction with the digital twin they can be very helpful for supporting the UN's SDGs, which we explore in the next section.

2.4 Digital Twin for Sustainable Development Goals

The UN encourages sustainable agriculture for Zero hunger, which is one of the goals of SDGs to end hunger and sustain food security with improvement in nutrition [2]. Food security can be nourished by implementing smart farming, mainly in urban areas [21]. In urban areas, almost half percent of the population resides [22], due to which cultivable land is insufficient. To tackle this problem, [21] encourages vertical smart farming in cities. Vertical smart farming can generate food crops in a larger amount in a small space with the support of agricultural technologies [23, 24]. introduced a research design paradigm in digital twins for vertical farming for two reasons: food productivity and resource utilization. This means the production optimization is expected to be carried out in a small area with multiple layers of land surfaces. In addition to that, it can be subsidized to save energy and utilize water, as climate change has driven water shortage. [11] explains how sensor data are processed using DL algorithms to detect nutrient deficiencies in plants.

The use of wind turbines and solar panels in vertical farming reduces carbon footprint emissions [21]. In rural regions, solar energy can be used for irrigation, which reduces the use of diesel for generating electricity [25]. The extreme use of non-renewable sources of energy has caused environmental pollution. The exploitation of renewable energy sources like solar energy, wind energy, etc. is accelerating in agricultural scenarios [26]. However, solar energy is considered the best option as it is the most abundant renewable source of energy [27]. Al-Ali et al. [28] and Chieochan et al. [29] are examples of how proper solar energy cells are installed according to the calculation for smart farm irrigation. The use of a photovoltaic system in a rural farm contributes to SDG 7: Affordable and clean energy. This scenario is seen in a work implemented in Portugal with 5527 kg of CO_2 savings and energy consumption reduced by 83.27% from an investment of €32,434 in 8 years of payback time [30].

The research works from Table 1 also contribute to sustainability with optimal production of food that will try to sustain food security by constant monitoring and supplying resources [9, 12]. monitor water content so that excess water will not be used. [11, 12] can be used for crop treatment and nutrient improvement.

In the next section, we study a digital twin paradigm that is implemented based on the theme we discussed earlier. This model attempts to create a digital twin farming for sustainable development.

3 A Case Study on Zero-Energy Farm Using Digital Twin Paradigm

With the concept implemented from recent studies in the field of agriculture, we study an ongoing project on a digital twin model which includes a zero-energy farm with a drip irrigation system (see Fig. 7). In [31], a zero-energy building is defined as a building that uses renewable energy to cover the building's needs. This model

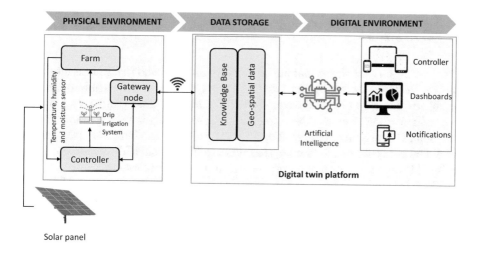

Fig. 7 Digital twin model for a zero-energy farm

implements the same idea on a farm. For sustainable agriculture, drip irrigation systems play a vital role by slowly delivering water directly to the root system. This reduces the loss of water from flooding and evaporation. The use of water is reduced by 40% more than traditional methods [32]. However, the IoT-enabled drip irrigation system was not explored till [33] introduced the concept to reduce excess water use and is very admired in recent years.

A 3KW solar panel is installed on a farm of 1200-square-meter area in Chiang Rai province, Thailand. The temperature, humidity, and moisture sensor will measure the environmental conditions which are immediately stored in a knowledge base through a gateway node with the help of Wi-Fi. LoRaWAN is used to connect devices to the Internet. The information on the physical environment from the sensors and geospatial data to locate the features in the farm is in storage. In a digital environment, the required data is accessed, and AI algorithms are responsible for decision-making with the knowledge to choose control strategies for optimizing production. Farmers and experts can access the information from the application to monitor the dashboard for visualization. It is also possible to control the devices on the farm manually or by setting the timer or automatically depending upon soil and weather conditions. Users will receive the notification on a condition in the physical environment recorded by the sensors.

The digital twin model is more interactive with the implementation of geospatial data. Users can easily access the location information of sensors and machinery equipment. The drip irrigation system minimizes the water runoff by delivering water slowly and uniformly. This leads to the efficient consumption of water and efficient production of crops which will contribute to SDG 2: Zero hunger. Moreover, the use of solar panels will subsidize SDG 7: Affordable and clean energy, as discussed in the above section. Most importantly, farmers or users can have remote access to the physical environment in real time.

4   Conclusion

This review chapter shows that the implementation of digital twins in farming will help in remotely accessing the farm to monitor the situation, possibly occurring problems like crop diseases, water content, the temperature in farms, etc. to take immediate action to protect the crop. According to the requirements, resources can be supplied in real time. The result of this review chapter also indicates that research in digital twin concepts in farming is exploited a little amount. The advancement in this field with more research is very important for efficient farming.

From the studies in this domain, we can acquire that the involvement of IoT sensors, AI, and cloud services has increased the possibilities of improving farming with automation and proper decision-making. The simulation software and application in the digital environment can help the limited number of farmers monitor and take instantaneous measures on a farm. This can help farmers to monitor a large scale of land and preserve crops and livestock. The implementation of photovoltaic systems can assist in some of the SDGs. Digital twin, overall, can have a huge contribution toward sustainable agriculture and the UN's SDGs. However, there is less research on the digital twin for the agricultural domain. There need to be more attempts for the advancement of technologies and ideas for more productivity in food for a sustainable future.

Acknowledgments This work was partially supported by the Thammasat University Research fund under the TSRI, Contract No. TUFF 19/2564 and TUFF24/2565, for the project of "AI Ready City Networking in RUN," based on the RUN Digital Cluster collaboration scheme and Center of Excellence in Intelligent Informatics, Speech and Language Technology, and Service Innovation (CILS), Thammasat University.

References

1. da Silva, J.G.: Feeding the world sustainably. UN Chron. **49**(2), 15–17 (2012)
2. United Nations: Investing in the SDGs: An Action Plan. United Nations Conference on Trade and Development, Geneva (2014)
3. Chakraborty, S., Newton, A.C.: Climate change, plant diseases and food security: an overview. Plant Pathol. **60**(1), 2–4 (2011)
4. Klomp, J., Hoogezand, B.: Natural disasters and agricultural protection: a panel data analysis. World Dev. **104**, 404–417 (2018)
5. Friess, P., Riemenschneider, R.: IoT ecosystems implementing smart technologies to drive innovation for future growth and development. Digit. Indust. **49**(2), 5–13 (2016)
6. Nasirahmadi, A., Hensel, O.: Toward the next generation of digitization in agriculture based on digital twin paradigm. Sensors. **22**(2), 498 (2022)
7. Jian, H., Qin, S., Fu, J., Zhang, J., Ding, G.: How to model and implement connections between physical and virtual models for digital twin application. J. Manuf. Syst. **58**, 36–51 (2021)
8. Kritzinger, W., Karner, M., Traar, G., Henjes, J., Sihn, W.: Digital twin in manufacturing: a categorical literature review and classification. IFAC-PapersOnLine. **51**(11), 1016–1022 (2018)
9. Alves, R.G., Souza, G., Maia, R.F., Tran, A.L.H., Kamienski, C., Soininen, J.P., Aquino, P.T., Lima, F.: A digital twin for smart farming. In: 2019 IEEE Global Humanitarian Technology Conference (GHTC), pp. 1–4. IEEE, Seattle (2019)
10. Jo, S.K., Park, D.H., Park, H., Kim, S.H.: Smart livestock farms using digital twin: feasibility study. In: 2018 International Conference on Information and Communication Technology Convergence (ICTC), pp. 1461–1463. IEEE, Jeju Island (2018)
11. Angin, P., Anisi, M.H., Göksel, F., Gürsoy, C., Büyükgülcü, A.: AgriLoRa: a digital twin framework for smart agriculture. J. Wirel. Mob. Netw. Ubiquit. Comput. Depend. Appl. **11**(4), 77–96 (2020)
12. Chaux, J.D., Sanchez-Londono, D., Barbieri, G.: A digital twin architecture to optimize productivity within controlled environment agriculture. Appl. Sci. **11**(19), 8875 (2021)
13. Zhang, Y., Kacira, M.: Enhancing resource use efficiency in plant factory. In: XXX International Horticultural Congress IHC2018: III International Symposium on Innovation and New Technologies in Protected 1271, pp. 307–314 (2018)
14. Bai, Y., Gao, J.: Optimization of the nitrogen fertilizer schedule of maize under drip irrigation in Jilin, China, based on DSSAT and GA. Agric. Water Manag. **244**, 106555 (2021)
15. Verdouw, C., Tekinerdogan, B., Beulens, A., Wolfert, S.: Digital twins in smart farming. Agric. Syst. **189**, 103046 (2021)
16. Muralidharan, S., Yoo, B., Ko, H.: Designing a semantic digital twin model for IoT. In: 2020 IEEE International Conference on Consumer Electronics (ICCE), pp. 1–2. IEEE (2020)
17. EnergyPlus: https://energyplus.net. Last accessed 20 Mar 2022

18. Hoogenboom, G., Porter, C.H., Shelia, V., Boote, K.J., Singh, U., White, J.W., Hunt, L.A., Ogoshi, R., Lizaso, J.I., Koo, J., Asseng, S., Singels, A., Moreno, L.P., Jones, J.W.: Decision Support System for Agrotechnology Transfer (DSSAT) Version 4.7.5. DSSAT Foundation, Gainesville. https://DSSAT.net (2019)
19. FarmBot: https://farm.bot/blogs/news/farmbot-digital-twin. Last accessed 20 Mar 2022
20. Murdyantoro, B., Sukma, D., Atmaja, E., Rachmat, H.: Application design of farmbot based on internet of things (IoT). Int. J. Adv. Sci. Eng. Inf. Technol. **9**(4), 1163–1170 (2019)
21. Musa, S.F.P.D., Basir, K.H.: Smart farming: towards a sustainable agri-food system. Br. Food J. (2021). https://doi.org/10.1108/BFJ-03-2021-0325
22. World Bank: Nutrition smart agriculture: when good nutrition is good business. https://www.worldbank.org/en/topic/agriculture/publication/nutrition-smart-agriculture-whengood-nutrition-is-good-business. Last accessed 20 Mar 2022
23. Saad, M.H.M., Hamdan, N.M., Sarker, M.R.: State of the art of urban smart vertical farming automation system: advanced topologies, issues and recommendations. Electronics. **10**(1422), 1–40 (2021)
24. Monteiro, J., Barata, J., Veloso, M., Veloso, L., Nunes, J.: Towards sustainable digital twins for vertical farming. In: 2018 Thirteenth International Conference on Digital Information Management (ICDIM), pp. 234–239. IEEE (2018)
25. Aroonsrimorakot, S., Laiphrakpam, M., Paisantanakij, W.: Solar panel energy technology for sustainable agriculture farming: a review. Int. J. Agric. Technol. **16**(3), 553–562 (2020)
26. Ali, S.M., Dash, N., Pradhan, A.: Role of renewable energy on agriculture. Int. J. Eng. Sci. Technol. **4**(1), 51–57 (2012)
27. Panwar, N.L., Kaushik, S.C., Kothari, S.: Role of renewable energy sources in environmental protection: a review. Renew. Sust. Energ. Rev. **15**(3), 1513–1524 (2011)
28. Al-Ali, A.R., Al Nabulsi, A., Mukhopadhyay, S., Awal, M.S., Fernandes, S., Ailabouni, K.: IoT-solar energy powered smart farm irrigation system. Journal of electronic. Sci. Technol. **17**(4), 100017 (2019)
29. Chieochan, O., Saokaew, A., Boonchieng, E.: Internet of things (IOT) for smart solar energy: a case study of the smart farm at Maejo University. In: 2017 International Conference on Control, Automation and Information Sciences (ICCAIS), pp. 262–267. IEEE (2017)
30. Pereira, F., Caetano, N.S., Felgueiras, C.: Increasing energy efficiency with a smart farm—an economic evaluation. Energy Rep. **8**, 454–461 (2022)
31. Torcellini, P., Pless, S., Deru, M., Crawley, D.: Zero Energy Buildings: a Critical Look at the Definition (No. NREL/CP-550-39833). National Renewable Energy Lab (NREL), Golden (2006)
32. Ranjan, S., Sow, S.: Drip irrigation system for sustainable agriculture. Front. Life Sci. **I**, 137
33. Jain, R.K., Gupta, B., Ansari, M., Ray, P.P.: IOT enabled smart drip irrigation system using web/android applications. In: 2020 11th International Conference on Computing, Communication and Networking Technologies (ICCCNT), pp. 1–6. IEEE (2020)

A Cross-comparison Between Rice Crop Monitoring Systems: GISTDA and International Asian Harvest mOnitoring System for Rice (INAHOR): JAXA

Kanjana Koedkurang, Patiwet Chalearmpong, Matawee Srisawat, and Panu Nuangjumnong

Abstract Rice is the most important crop governing food security in Asia. Rice crop monitoring is an important activity for crop management. While space technology and geo-informatics technologies are now being used in a wide range of applications, Geo-Informatics and Space Technology Development Agency (Public Organ-ization) (GISTDA) has developed Thailand Rice Crop Monitoring System (TRMS) to monitor rice cultivation areas. The system is to the present based on passive and active satellite sensors (Landsat-8, Sentinel-2, Thaichote, Radarsat-2, and Sentinel-1). The aims are to report the status and timelines of rice cultivation from the initial step until harvest seasons to estimate rice production for Thai agricultural management.

Japan Aerospace Exploration Agency (JAXA) developed the INAHOR (International Asian Harvest mOnitoring system for Rice) system which is a tracking information system to check the rice harvest in Asia. It was developed to assess rice harvesting areas using ALOS-2 satellites to refine INAHOR by operating in Southeast Asia. This reduces the limitation of the use of optical satellite data in the case of clouds during the rainy season which is the period of rice cultivation; thus, the rice cultivation can be monitored throughout the cultivation period and in every planting cycle.

This study aimed to evaluate the potential and suitability for use of the TRMS-GISTDA and INAHOR-JAXA by comparing the results of the rice planting areas of the TRMS from GISTDA and the harvesting area of the INAHO from JAXA with the statistical data from the Office of Agricultural Economics because both systems have advantages and limits. And their approaches can be used to analyze and evalu-

K. Koedkurang (✉) · P. Chalearmpong · M. Srisawat · P. Nuangjumnong
GISTDA 120 The Government Complex Commemorating His Majesty The King's 80 th Birthday Anniversary, 5th December, B.E.2550(2007), Lak Si Bangkok, Thailand
e-mail: kanjana@gistda.or.th

W. Boonpook et al. (eds.), *Applied Geography and Geoinformatics for Sustainable Development*, Springer Geography,
https://doi.org/10.1007/978-3-031-16217-6_10

ate the rice-growing areas in other areas such as in the ASEAN region. It can be utilized to improve the efficiency of rice growing in the country.

Keywords Thailand Rice Crop Monitoring System (TRMS) · International Asian Harvest mOnitoring system for Rice (INAHOR) · Rice planting areas

1   Introduction

Rice is among the most important foods as it not only feeds half of the world's population but plays a vital part in national economies. Asia accounts for about 90% of the global rice production and consumption – which means that rice production in Asia is instrumental to the world's food security [1]. Southeast Asia makes up for approximately 30% of the global rice production areas, and according to the data on exports in 2021, Thailand's rice exports rank third behind India and Vietnam [2, 3].

Monitoring rice planting areas is key to the projection and management of the crop: the data collected can be useful in decision-making, planning, and policy formulation at national and regional levels. At present, space and geo-information technologies have been adopted to facilitate work in various fields, given that the data collected using these technologies cover a wide range of areas and that they have been continuously recorded for quite some time. When these data are put into analyses together with other data using advanced computers and processing techniques, they make it possible to manage resources and the environment and, specifically, to track and monitor rice production areas, measure land fertility, and forecast crop yields in a more sustainable fashion and with greater efficiency.

Even though GISTDA and JAXA have in common the rice crop monitoring system feature based on the satellite data, they are different in various terms such as data type, method, and area coverage. To be specific, GISTDA relies heavily on the data from optical satellites and sporadically on additional active data, in measuring rice production areas in Thailand. The data are collected from the first step of rice production until harvest using the "unsupervised" method, K-mean, and the hidden Markov model. In contrast, INAHOR (International Asian Harvest mOnitoring system for Rice) from JAXA is a tracking system that monitors rice harvesting in Asia – using the machine learning feature called random forest. The feature is based on the ALOS-2 data from L-Band (PALSAR-2) wavelength in measuring rice harvesting areas in Asia, which helps to avoid limitations of using the data from optical satellites in the rainy and cloudy season when rice is being grown. This technique makes it possible to monitor throughout every stage and round of rice production.

Because both TRMS-GISTDA and INAHOR-JAXA have different advantages and limitations, this study aimed to examine the potential and suitability of the two systems for implementation by comparing the rice planting and harvesting areas in Thailand from TRMS-GISTDA and INAHOR-JAXA, with the statistics data from the Office of Agricultural Economics (OAE) to identify how to apply them to improve the monitoring of rice cultivation and to be more effective.

2   Objective

1. To compare the results of rice cultivation and harvest area between the TRMS-GISTDA and the INAHOR-JAXA
2. To examine the potential and suitability of both the TRMS-GISTDA and the INAHOR-JAXA system

3   Thailand Rice Crop Monitoring System (TRMS): GISTDA

GISTDA has developed a model and a processing system for monitoring rice cultivation areas from satellite data. The monitoring system of rice cultivation in Thailand from satellite data or the Thailand Rice Crop Monitoring System is updated every 2 weeks. The state classification of rice is divided into 2-, 4-, 6-, 8-, 10-, 12-, 14-, and 16-week rice, respectively, from 2014 to the present. TRMS uses both optical and active remote sensing satellte as Landsat-8, Sentinel-2, Thaichote, Sentinel-1 and Radarsat-2 for monitoring the situation of rice cultivation from the start of the crop to the harvest to be able to calculate the amount of production in each area which this system has been made into an online system for providing an online plantation area mapping service (https://rice.gistda.or.th/) that is made accessible to related from both local and national agencies to allowing them to use the data to improve their decisions in land management [4].

Monitoring of rice cultivation areas with the TRMS system is divided into three main parts.

3.1   Rice Crop Monitoring from Optical Satellite Data

The rice planting areas are interpreted from optical satellites (Landsat-8, Thaichote, and Sentinel-2) using the K-means unsupervised classification method. Then, the current stages of rice plants are labeled by experts as Nothing, Growing, Mature, and Harvest. This enables not only the ricc planting areas to be identified but also the estimation of the current rice state and expected harvest days to be made. In general, the rice production cycle starts with "Growing" into "Mature" and "Harvest" and ends with "Nothing" as illustrated in the ideal NDVI graph (Fig. 1, left) [5]. With four stages in a cycle, the possibility graph depicting the transition from one stage to another can be seen in the hidden Markov model (Fig. 1, right) [5–7],

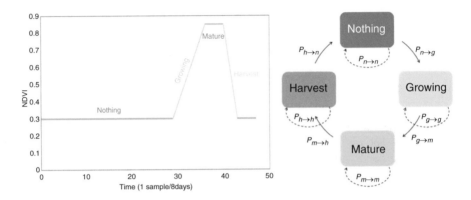

Fig. 1 An ideal NDVI graph at different growing states and at different times (left) and hidden Markov model of NDVI in all four growing states (right)

3.2 Rice Crop Monitoring from Active Satellite Data

The rice planting areas are identified from active remote sensing satellites: Radarsat-8 (R: HH, G: HV, B: HH-HV) and Sentinel-1A (R: VV, G: VH, B: VV-VH) using the K-means unsupervised classification method like in the previous step. Then, instead, the current stages of rice plants are labeled by experts as "Growing" or "Not Growing."

3.3 Rice Crop Monitoring from Multi-resolution

Given that the TRMS-GISTDA takes into several the data from different satellites with active or passive sensors – Landsat-8, Thaichote, Sentinel-2, Radarsat-2, and Sentinel-1, for instance – it is necessary to process the data from all those satellites together to reduce discrepancies in pixel resolution to reach an output with a single resolution by choosing the resolution of the result to be served at 50 × 50 m. The process of rice crop monitoring in Thailand is shown in Fig. 2.

4 JAXA's INAHOR (International Asian Harvest mOnitoring System for Rice)

INAHOR-JAXA is a tracking information system that checks the rice harvest in Asia. Its main function is to provide a rice planted area map (including the growing stages of classification) and a rice planted area and production (needs yield information) [8, 9] using Synthetic Aperture Radar data in the L-Band wavelength

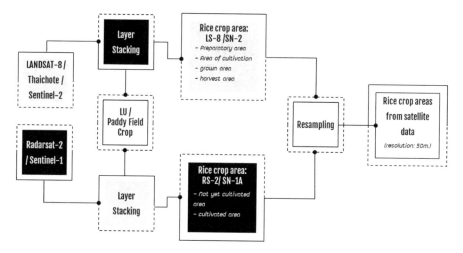

Fig. 2 Workflow of TRMS-GISTDA process

- Paddy rice area has "Inundation" and "Vegetative" stages.

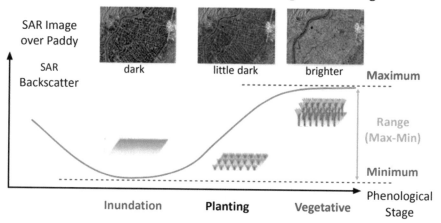

Fig. 3 Basic concept for identifying rice planted area

(PALSAR-2) from the ALOS-2 satellite, which reduces the limitation of the use of satellite data in optical systems in terms of clouds during the rainy season which is the period when rice is planted. Thus, the rice cultivation can be monitored throughout the cultivation period and in every planting cycle.

The INAHOR-JAXA uses machine learning: random forest to identify rice crop areas. The algorithm works by analyzing the time series data, given that the backscattering data targeting rice crop areas will be "dark" during rice-growing stages, and these data will be "bright" after harvest, as shown in Fig. 3 [10]. This feature

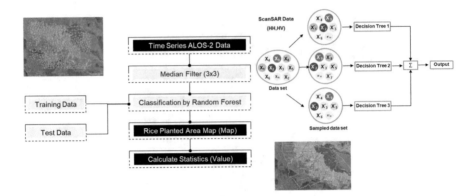

Fig. 4 Workflow of INAHOR process

requires training data in various patterns to teach the system, allowing it to accurately recognize and categorize rice crop areas and each crop stage in the areas. How this works can be found in Fig. 4.

5   Cross-Comparison Between TRMS and INAHOR with OAE

In terms of evaluating the efficiency of each system, the statistical data from the Office of Agricultural Economics (OAE) was used as the base. OAE collects statistical data for rice planting and harvesting areas over Thailand which were collected at the subdistrict, district, and provincial levels every month.

The cultivation area from the TRMS-GISTDA system and the harvest area from the INAHOR-JAXA system during the major-season rice in 2018 and 2019 will be analyzed and calculated at the provincial, regional, and national levels. Then it will be compared with the statistical data of the cultivated area and the harvested area by the province of the OAE during the same period as shown in Fig. 5.

6   Results

In this study, the rice planting area from TRMS-GISTDA (Fig. 6) and the rice harvesting area from INAHOR- JAXA (Fig. 7) based on major-season rice in 2018 and 2019 were compared to the statistical data of rice planting and harvesting areas from the OAE in the same period (Table 1 and Fig. 8). The results are as

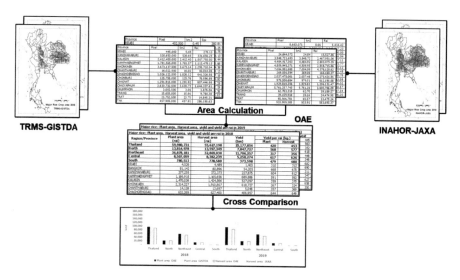

Fig. 5 Workflow of cross-comparison process

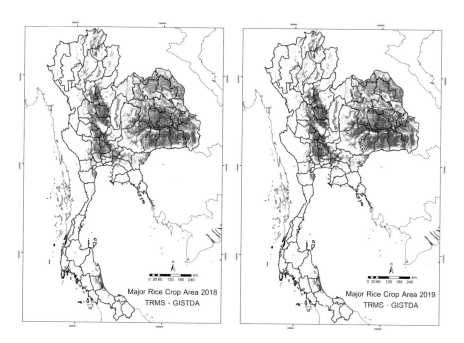

Fig. 6 Rice planting area from TRMS-GISTDA in 2018 (left) and 2019 (right)

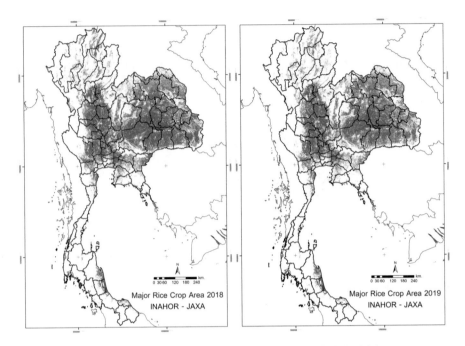

Fig. 7 Rice harvesting area from INAHOR-JAXA in 2018 (left) and 2019 (right)

Table 1 The rice planting area from TRMS-GISTDA and the rice harvesting area from INAHOR-JAXA compared to the statistics data from OAE in 2018 and 2019

Year	Area	Plant area (km²)		Harvest area (km²)	
		OAE	GISTDA	OAE	JAXA
2018	Thailand	95,969.17	97,220.70	89,003.52	162,523.34
	North	22,103.96	21,959.78	1755.76	35,090.49
	Northeast	9005.09	59,941.09	52,590.45	91,623.79
	Central	3611.29	14,540.89	13,411.58	2680.70
	South	1248.82	778.95	1245.73	3128.35
2019	Thailand	97,915.41	96,747.58	86,573.24	137,799.69
	North	22,617.56	21,347.65	21,385.86	37,648.84
	Northeast	60,478.65	60,236.21	51,236.40	36,841.22
	Central	13,584.03	14,049.41	12,783.50	33,927.82
	South	1235.18	1114.32	1167.48	29,381.81

follows. Overall, at the national level, the result from TRMS-GISTDA is close to the data from OAE, especially in northern and northeastern regions but with the southern region showing the largest discrepancy between the two databases. Meanwhile, the result from INAHOR-JAXA shows an overestimate of the OAE data, especially in the southern region.

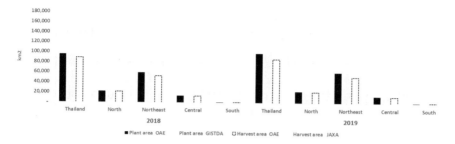

Fig. 8 The chart of planting area from TRMS-GISTDA, the harvesting area from INAHOR-JAXA, and the statistics data from OAE in 2018 and 2019

7   Conclusion and Discussions

The study of the rice crop monitoring system and comparing the results of the rice planting and harvesting areas observed by different methods found that the result from TRMS-GISTDA is close to the OAE data than the INAHOR-JAXA at the national level. When considering the Thailand region level, the rice cultivation and harvesting areas from both systems in the southern part of Thailand were different from the OAE data than other regions.

Considering the advantages, limitations, and suitability of each system, the INAHOR-JAXA is a tracking information system that checks the rice harvest at the regional level such as ASEAN based on the data from ALOS-2 and a certain amount of training data. If the training data are sufficient and well represented of how and what rice is grown in each area or region, the analysis will invariably become more accurate. Also, if the analysis is to include a more comprehensive set of data (collect-ed over two years), it may potentially lead to a reliable model or formula that will improve the estimate of rice harvesting areas, which will enable the forecasts of annual crop yields in each region to be more accurate.

For the TRMS-GISTDA is a rice crop monitoring system in Thailand, relying largely upon data from satellites together with land use data and experienced experts in its analysis and identification of each rice planting area, as well as the growing stage of the rice plants being grown throughout the cycle. Even though this system shows the results that are similar to the data collected by OAE, adopting it in a different country requires the same type of data that is available in Thailand as well as experts in the given field, which may take a lengthy period to develop to ensure the reliability of the result.

In the cross-comparison results of the rice planting and harvesting areas between TRMS-GISTDA and INAHOR-JAXA with the OAE data, substantial differences have been found, which can be caused by various factors as the data resolution from TRMS-GISTDA is 50 m, whereas the data resolution of INAHOR-JAXA is 100 m, which may have led to the classifying results of more rice crop areas than TRMS-GISTDA and OAE. In Thailand, the rice crop cycles start in different months, depending on the region, and the southern part has the most different rice-growing season among all the regions. This resulted in the southern region showing the largest data discrepancies between both rice crop monitoring systems. It thus may be necessary to analyze and process the data from both systems separately, considering that the INAHOR-JAXA, which employs machine learning, may also need a greater amount of training data that better represent the unique characteristics of rice production in each region. As rice production cycles start in different months for different regions and the length of each cycle varies depending on the rice variety, the harvest seasons arrive at different months across regions. This means that when only the total amounts from each system are considered, the results can easily be inaccurate.

Acknowledgment We sincerely thank the JAXA-SAFE team for sharing the rice cultivation area data from the INAHOR system and permission to use it in this research. And we are grateful to work on the SAFE project together.

References

1. Han, J., Zhang, Z., Luo, Y., Cao, J., Zhang, L., Cheng, F., Zhuang, H., Zhang, J.: AsiaRiceMap10m: high-resolution annual paddy rice maps for Southeast and Northeast Asia from 2017 to 2019. Earth System Science Data (2021)
2. Springnews: ดับฝันส่งออกข้าวไทยร่วงลงอันดับ4ของโลก ! คู่แข่งมาแรงแซงทางโค้ง, NATION DIGITAL CONTENT CO., LTD., 09 March 2022. [Online]. Available: https://www.springnews.co.th/news/806740#google_vignette. Accessed 30 Mar 2022
3. Shahbandeh, M.: Principal rice exporting countries worldwide in 2021/2022. Statista, 31 January 2022. [Online]. Available: https://www.statista.com/statistics/255947/top-rice-exporting-countries-worldwide-2011/. Accessed 30 Mar 2022
4. Panu: Economic Crop Monitoring and Yield Forecasting with Geospatial Technology. Geo-Informatics and Space Technology Development Agency (Public Organization), Bangkok (2017)
5. Suwannachatkul, S., Kasetkasem, T., Chumkesornkulkit, K., Rakwatin, P., Chanwimaluang, T., Kumazawa, I.: Rice cultivation and harvest date identification based on a hidden markov model. In 11th International Conference on Electrical Engineering/Electronics, Computer, Telecommunications and Information Technology (ECTI-CON), Nakhon Ratchasima, Thailand (2014)
6. Siachalou, S., Mallinis, G., Tsakiri-Strati, M.: A hidden Markov models approach for crop classification: linking crop phenology to time series of multi-sensor remote sensing data. Remote Sens. **7**(14), 3633–3650 (2015)
7. Suwannachatkul, S., Kasetkasem, T., Chumkesornkulkit, K., Rakwatin, P., Chanwimaluang, T., Kumazawa, I.: Rice cultivation and harvest date identification based on a hidden Markov model. In 014 11th International Conference on Electrical Engineering/Electronics, Computer. Telecommunications and Information Technology (ECTI-CON), Nakhon Ratchasima, Thailand (2014)
8. Okumura, T.: Asia-RiCE: Rice Crop Es2ma2on and Monitoring (the role of PALSAR-2)." RESTEC, 18 January 2017. [Online]. Available: https://www.eorc.jaxa.jp/ALOS/en/kyoto/jan2017_kc23/pdf/1-05_KC23_Okumura.pdf. Accessed 25 Mar 2022
9. Oyoshi, K., Tomiyama, N., Okumura, T., Sobue, S., Sato, J.: Mapping rice-planted areas using time-series synthetic aperture radar data for the Asia-RiCE activity. Paddy Water Environ. **14**, 463–472 (2016)
10. Hamamoto, K., Sobue, S., Oyoshi, K.: JAXA's Mission Updates and Agricultural Applications. Japan Aerospace Exploration Agency (JAXA), 28–30 May 2018. [Online]. Available: https://lcluc.umd.edu/sites/default/files/Hamamoto_0.pdf. Accessed 25 Mar 2022.

Evaluation MODIS and Sentinel-2 Data for Detecting Crop Residue Burned Area

Chanarun Saisaward and Sarawut Ninsawat

Abstract Crop residue burning is one of the major sources of air pollution problems. Various studies have been focused on burned area detection by using 500-m MODIS (MCD64A1) because of the high temporal and ready-to-use product. MCD64A1 data was widely used and well detected in the wildfire area. But the underestimated results in agricultural fields were found in many studies. Because its capabilities of coarse spatial resolution cannot clearly detect small fire patches. This study aims to evaluate the MODIS burned area product and Sentinel-2 burned area for crop residue burning. Higher spatial of burned areas were analyzed by Normalized Burn Ratio (NBR) from Sentinel-2 to compare with MODIS burned area product. The burned areas were overlayed with land uses and sugarcane plots in the production season 2018/2019 in Kalasin province and two districts of Udon Thani province. The estimation of the sugarcane harvesting period was done by NDVI derived from 15-day composite of Sentinel-2 imageries. The result showed that the total burned from Sentinel-2 had around 11 times higher than MCD64A1. The majority of detected burned areas of both sensors were detected in the paddy field. Additional, MCD64A1 has better performace to identify the forest burn area than the Sentinel-2 where as it is not suitable to detect the field crops which has relative smaller size. The study reports that most sugarcane in the study area was harvested in the first half of January 2019 with the cutting of fresh sugarcane, while burned harvest was mostly found in the first half of April 2019. Therefore, this study could fill the gap of present knowledge about crop residue burning over the region.

Keywords Burned area · Normalized Burn Ratio (NBR) · Crop residue burning · MCD64A1 · Sentinel-2

C. Saisaward · S. Ninsawat (✉)
Remote Sensing and GIS FoS, School of Engineering and Technology, Asian Institute of Technology, Pathumthani, Thailand
e-mail: sarawutn@ait.ac.th

© The Author(s), under exclusive license to Springer Nature Switzerland AG 2023
W. Boonpook et al. (eds.), *Applied Geography and Geoinformatics for Sustainable Development*, Springer Geography,
https://doi.org/10.1007/978-3-031-16217-6_11

1 Introduction

There is an upward trend in demand for food consumption worldwide resulting in the growth rate in agricultural production. In Thailand, agricultural areas dominated the most considerable amount of land, contributing over 24 million hectares or 46% of the total land, which mainly are paddy, sugarcane, and maize. Agricultural residue burning is a type of open biomass burning which is a general practice and is preferred by farmers as it is a rapid and low-cost technique for clearing land [1]. The preharvest burning residues of sugarcane are extremely hazardous and remove organic matter from the soil [2]. The MODIS fire products have been commonly used in many studies for characteristics of fire in forest areas [3–6], and the previous study found that the proportion of wildfire detected by MODIS in the forest and savanna/grassland burning provides a persistent and reliable method over large areas compared to other land uses [7, 8].

Hantson found that the MODIS hotspots are only 1.8% reliable to detect true burned areas [9]. Therefore, MODIS burned area products (MCD64A1) are used instead of fire hotspots in this study. However, most studies found that because of the coarse spatial resolution of MCD64A1, they are not likely able to detect small-burned patches in cropland, for example, burn patches below 120 ha in size [10]. Also, it found that MCD64A1 tend to underestimate the rice residue burning around 44% compare to the result of hybrid MCD64A1 and Landat data [11]. Kiadtikornthaweeyot adopted 30-m spatial resolution of Landsat-8 NBR for detecting burned areas [12]. The spatial resolution of 30 m makes this dataset good for detecting and analyzing the land after the fire. It is applicable to identify small-scale crop burned areas by overcoming limitations of MODIS, but it needs to trade off for its lower temporal resolution.

In this study, the burned area in agricultural and croplands is the main target feature, with additional focus on the burning of sugarcane which was one of the main crops in the study area. This study aims to evaluate the MODIS burned area product and Sentinel-2 burned area for crop residue burning with the research question that is the burned area product from the MODIS sensor shows the different results from Sentinel-2 burned area. In addition, the identification of the estimated harvesting period was analyzed and matched up with the sugarcane burned plots to indicate the burned method of each farm between preharvest burned and cutting of fresh sugarcane.

2 Material and Methods

2.1 Study Area

The study area was located in Kalasin province and two districts of Udon Thani province which is Wang Sam Mo and Si That district in the upper northeastern Thailand. The study areas were selected because there are a large number of

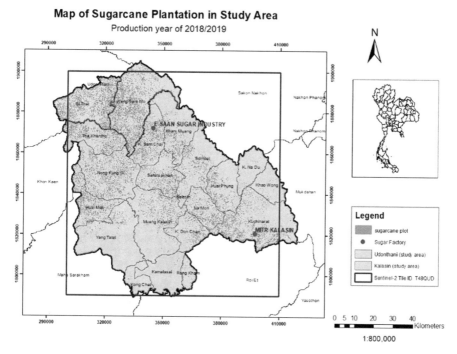

Fig. 1 Map of sugarcane plantation in the study area

sugarcane plots, sugarcane production, and yield compared to other provinces in northeastern Thailand during the production year 2018/2019. Moreover, all areas of these provinces were covered in the whole one scene of Sentinel-2 tile number T48QUD that was applicable to interpret and further do analysis (see Fig. 1).

2.2 Dataset

The daily MODIS MCD64A1 burned area product with 500 meter resolution was used in this study which was the improved version of burned area mapping to detect small burns. The products were downloaded from https://lpdaacsvc.cr.usgs.gov/ appeears covering the study period between November 2018 and April 2019. The algorithms employed a couple of burn vegetation indexes from surface reflectance bands 5 and 7 of MODIS imagery and 1-kilometer MODIS active fire observations as input data. This product provided high temporal resolution but had a limitation on the spatial resolution of 500 meters. The burned pixel was indicated by the burn date layer showing in Julian date (1-366). Meanwhile, the higher spatial resolution to indicate the burned area was employed by 20 meter NIR (band 8A) and SWIR (band 12) of Sentinel-2 using the Normalized Burn Ratio index. There was a total of 20 couples or 40 scenes of suitable Sentinel-2 scenes with cloud-free and suitable to process in this study. The sugarcane planting areas in each farm were indicated by sugarcane plot boundary obtained from the Office of the Cane and Sugar Board (OCSB) in the production year 2018/2019. There were 39,181 plots in the study area used for processing. Moreover, the land use in the year 2019 obtained from Land Development Department was employed to evaluate the burned area in each land use. A total of 31 different types of land use classes was found and reclassified into seven main classes. The study period of this study was November 2018–2019 because sugarcane is normally harvested in this period depending on the sowing time.

2.3 Methodology

The schematic diagram of the overall methodology is presented in Fig. 2. The first step, the evaluation between the burned area from MODIS (MCD64A1) burned area product and Sentinel-2, was performed by analyzing the total amount of area and the percentage of sugarcane burned area. The burned area from the processing was overlayed with land uses and sugarcane crop plots for indicating residue burning area.

    The verification plots from the field survey were collected at Kumphawapi and Nonsaat district in Udon Thani province with the estimation date and period of burn activity during December 2020 and January 2021. These plots were used to validate burned plots from Sentinel-2. The Sentinel-2 tile ID T48QTD was used to create

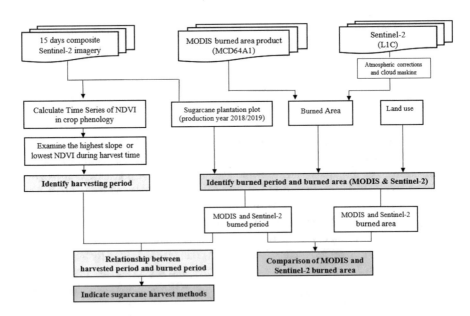

Fig. 2 Overall methodology

eight couples of NBR images from 14 December 2020 to 5 February 2021. The processing steps were done as same as the burned products from the study area. Moreover, unburned plots were collected from Google Earth Pro in the same study area and study period to assess the false negative of the burned results.

Subsequently, the objective of this part aimed to determine the relationship between sugarcane residue burning and sugarcane crop calendar by using 15-day composite of Sentinel-2 imagery. The process was done on the Google Earth Engine platform. The highest slope of NDVI during the burning period was then identified as the harvesting period. After analyzing the burned period and burned area from the first step, the relationship between both data was conducted to examine the method of harvesting sugarcane which has two types: (1) cutting of fresh sugarcane and (2) burned method.

2.3.1 Identification of Burned Area from MODIS

The MCD64A1 products were downloaded and projected into World Geodetic System (WGS) 1984 UTM zone 48N projected coordinate system. The number of daily burned pixels was indicated by each attribute shown in the Julian date of burn. The water area and unburned area were also specified. Since there was a temporal gap between both sensors, the daily burned pixels from MODIS were grouped and compiled to the total range based on the same range as Sentinel-2's acquisition dates to comparable.

2.3.2 Identification of Burned Area from Sentinel-2

To identify areas affected by fire from Sentinel-2, the near-infrared (NIR) and short-wave infrared (SWIR) were used to calculate Normalize Burn Ratio (NBR). The delta NBR (dNBR) is then obtained from the difference NBR between pair of images from Date1 and Date2 to estimate burn severity. Afterward, dNBR was then classified according to the United States Geological Survey (USGS) standard for burn severity assessment. Finally, to highlight burned areas, seven levels of the burn severity were then reclassified to burn and unburn pixels. High severity, moderate-high severity, and moderate-low severity are indicated to burn. On the other hand, low severity and post-fire high and low enhanced regrowth, also unburn class, were reclassed to unburned areas. There were 20 couples or 40 scenes of suitable Sentinel-2 imageries with the minimum cloud cover to compute dNBR that cover the burned period of six months from November 2018 to April 2019 in the study area. The equations for calculating NBR and dNBR are shown in Eqs. 1 and 2, respectively:

$$NBR = (NIR - \text{SWIR}) / (NIR + \text{SWIR}) \tag{1}$$

$$dNBR = NBR(\text{Date1}) - NBR(\text{Date1}) \tag{2}$$

Subsequently, the burned results were then overlayed with land used data and sugarcane crop plots to identify crop burned areas. The estimated burned period of each sugarcane farm was indicated from the acquisition date of couple Sentinel-2 imageries.

After identifying burned area from both datasets, the total burned area and percentage of burned area detected from both products were calculated in terms of spatial distribution of different land uses and burned area in sugarcane fields.

2.3.3 Identify Sugarcane Harvesting Period and Methods

Since there are attempts to reduce and control the sugarcane preharvest burning by government in order to reduce air pollution sources, this section aims to determine harvesting methods between cutting fresh sugarcane and the preharvest burning method of total sugarcane farms in study area. Firstly, the investigate harvest period of each sugarcane farm had to be analyzed due to the unavailable harvesting period in the sugarcane plot datasets. The harvesting period was identified from the time series of NDVI analyzed in Google Earth Engine platform using 15-day composite of Sentinel-2 time series imageries. The algorithm created random points in each sugarcane farm. The maximum NDVI value for the random points was extracted and the average NDVI for each farm was computed. The rule was the first lowest NDVI value during harvesting season (November to April) was identified as the harvesting period, because the NDVI phenology of harvested area will be dramatically dropped to the lowest point in the plantation cycle. Figure 3 showed one crop cycle of sugarcane time series of NDVI phenology which can estimate the harvested

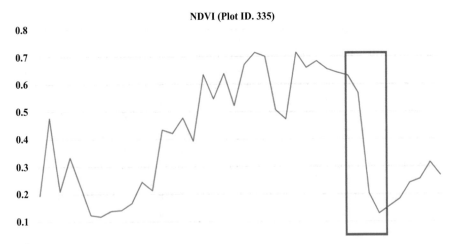

Fig. 3 Sugarcane NDVI phenology in one production year

period from the highest NDVI to the lowest around the second half of December to the first half of January.

Subsequently, on the account of the time series NDVI can monitor only the harvest period, not the harvest methods that farmers use. The estimated burn period information was pointed out by the acquisition date of Sentinel-2's burned area imageries to imply and match with the harvested period. Day of the year in Julian year of 2018–2019 was employed for specifying to set as the same time temporal of both datasets. The duration of harvest and burn was organized into 12 periods, each period consisting of 15 days from the first half of November, the second half of November, and until the second half of April. After labeled harvested period and burn period, all plots in both datasets were intersected. The relationship between burned and harvesting periods showed which plots had the same period of harvest and burn. Therefore, this could be implied that these plots use the burned method to preharvest sugarcane, while the plots that only harvest but do not detect burned in the same duration were implied to the fresh cutting method.

3 Results

3.1 Identification of Burned Area from MODIS

The result of total MODIS burned area from the number of pixels burned in the burn date layer during burning season 2018/2019 was 236 km^2 or 1152 pixels. This number was accounted for only 3% of the total study area. According to the total of just over 40,000 pixels or 8264.54 km^2, the products detected per-pixel burned area in November of around 616 pixels or 126 km^2, which had almost three times of burned pixels in January and February with 226 pixels (46 km^2) and 173 pixels (35 km^2), respectively. On the other hand, the burned pixels for December, March, and April had a very small number of under 50 pixels or less than 10 km^2.

In terms of the daily burned area from November to April (Fig. 4), the maximum of burned pixel was detected in mid-November, at the peak of approximately 35 and 26 km^2 on 13 and 15 November 2018. After that, a few burned detections were found from 16 November 2018 until the first week of January. Then, the number was high and fluctuated between few square kilometers to 6 square kilometers from mid-January until the end of the study period.

3.2 Identification of Burned Area from Sentinel-2

There are 20 couples or 40 scenes of suitable Sentinel-2 imageries with the minimum cloud cover to compute dNBR. According to Fig. 5, the 24th of January 2019 image was represented for the before image and the 29th of January 2019 was used for after burn activities in this period and then was used to be the before image for

Daily burned area derived from MCD64A1

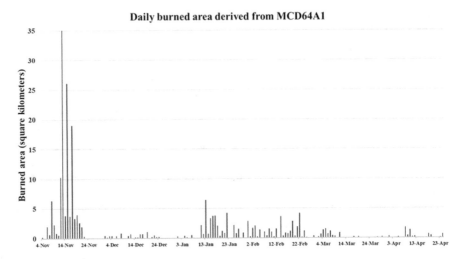

Fig. 4 MCD64A1 daily burned pixels

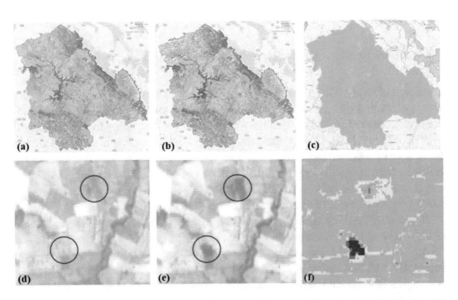

Fig. 5 Burn severity mapping of the 24–29 January 2019 dNBR image. (**a**) is before-fire (24 January 2019), (**b**) is the after-fire image (29 January 2019) after applying atmospheric correction, (**c**) is classified dNBR to severity classes, (**d**), (**e**), and (**f**) are zoom to their plot-burn level of (**a**), (**b**), and (**c**), respectively

the next period. The burn area are clearly identified and shown as darker pixels in Fig. 5(e), these pixels are estimated as different severity levels depending on the value of dNBR. Subsequently, the dNBR value was classified to the different levels of burn severity. Table 1 shows the amount of burned area classified by severity classes between 24 and 29 January 2019 in the study area. A maximum number of

Table 1 Burned area classified by severity classes of the 24–29 January 2019 dNBR image

Severity class	Area (km²)	Percentage (%)	Pixels	Reclass
High severity	1.73	0.02	192	Burn
Moderate-high severity	14.81	0.17	16,455	
Moderate-low severity	57.51	0.67	63,896	
Low severity	516.31	6.02	573,674	Unburn
Unburned	7303.75	85.19	8,115,276	
Enhanced regrowth, low	545.18	6.36	605,755	
Enhanced regrowth, high	133.95	1.56	148,836	
Total	8573.23	100	9,524,084	

land change detection was observed in the low severity class, accounting for over 500 km² (6.02%), while the other classes had a few percentages of burn severity detection. The percentage for unburned land detection had a vast majority (85%) of the total. After reclassifying the severity levels into burn or unburn pixels, it is observed that the burned area is highly less compared to the unburned area in this period. Only lower than 1% of the total area was observed to be the burned area.

3.3 Comparison of the Burned Area from MODIS and Sentinel-2

The comparison of the MODIS MCD64A1 burned product and the NBR from Sentinel-2 has been carried out during the crop burning season in the production year of 2018–2019. Based on the total of 8147 km² in the study area, the burned result was 221 km² (2.72%) of the overall burned area assessed by the MODIS MCD64A1. In comparison, the total burned areas derived from Sentinel-2 were approximately 2430 km² (30%). Therefore, it was found that the total burned from Sentinel-2 had around 11 times higher than MCD64A1. The peak burning from both sensors was found between 5 November and 25 November 2018, accounting for 125 km² and 1211 km² from MCD64A1 and Sentinel-2 (see Fig. 6).

Based on the MCD64A1 product, there is an improvement in the ability to detect small fires in this version. However, its resolution of 500-m per pixel was too coarse to detect fire in the agricultural area in this region. As shown in Fig. 7, taking the period date of 24–29 January 2019 as an example, burned area derived from Sentinel-2 (yellow pixel) was more appropriate with the scale-size farm plot while one pixel of burned MCD64A1 (pink pixel) covered many farms. As a result, MCD64A1 detected only a big cluster of burned areas from Sentinel-2.

In terms of the proportion of burned area in different land uses, the burned area from both products was overlaid with land use data over the study area from LDD to visualize the proportion of burned area in each land use. The MODIS burned results showed that the proportion of burned area detected in the paddy field far exceeded that of the other land uses: forest, field crop, agricultural, and

Comparison of the burned area from MODIS and Sentinel-2

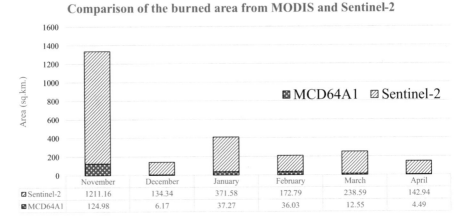

	November	December	January	February	March	April
◿ Sentinel-2	1211.16	134.34	371.58	172.79	238.59	142.94
◼ MCD64A1	124.98	6.17	37.27	36.03	12.55	4.49

Fig. 6 Comparison of the burned area from MODIS and Sentinel-2

Comparison of burn pixel from MCD64A1 and Sentinel-2 dNBR

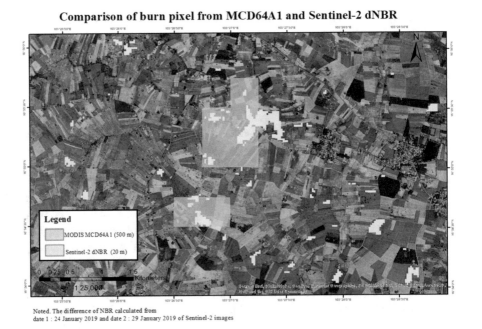

Noted. The difference of NBR calculated from
date 1 : 24 January 2019 and date 2 : 29 January 2019 of Sentinel-2 images

Fig. 7 Comparison of burn pixel from MCD64A1 and Sentinel-2 dNBR

miscellaneous, standing at 148 km^2 compared to 35 km^2 in the forest, 24 km^2 in field crop, 7 km^2 in agricultural, and only a few sq.km. in miscellaneous land (Fig. 8). Moving on to burned area derived by Sentinel-2, the highest amount of burned area also occurred in the paddy field (1389 km^2), but a prominent figure of 1049 km^2 (76%) showed at only in November and dramatically dropped to few percentages of burned in the paddy field on other months, while the burned area in field crop

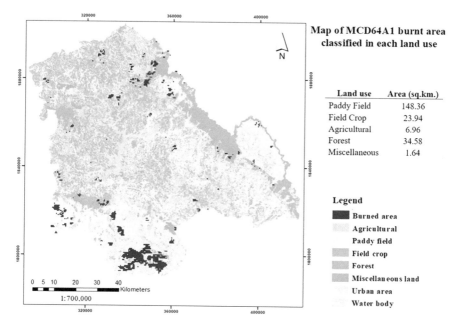

Fig. 8 Map of MCD64A1 burned area classified in each land use

fluctuated and remained higher area than paddy field and other land use except in April which the figure in paddy field was higher than in field crop. Burned area detected in the forest had almost 11 times less than in field crop, at around 56 km², which showed the different trends of the higher burned area from MCD64A1.

3.4 Sugarcane Residue Burning

In production season 2018/2019, the amount of burned sugarcane was identified by overlaying burning from Sentinel-2 with sugarcane plantation areas provided by OCSB. The total of burned sugarcane was 283 km² (3.47% of total area) from November 2018 to April 2019. This number accounted for 11.35% of the overall burned area throughout the study area. In terms of temporal (Fig. 9), sugarcane burned data from each period of Sentinel-2 were grouped by month to study the distribution pattern. The largest number of burned area in sugarcane plantation occurrence covered with the same figures of 64 km² in January and March, followed by November, February, and December, amounting to 59 km², 47 km², and 35 km², respectively, while April had the least number of burned sugarcane.

Considering the percentage of burned sugarcane over total sugarcane plantation area in each sub-district, Nong Suang (39.96%), Na Kham (35.33%), and Samran Tai (34.93%) are the largest proportion (Fig. 10).

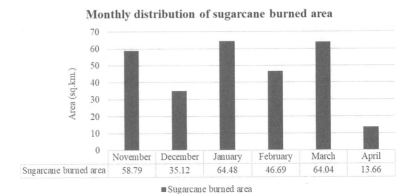

Monthly distribution of sugarcane burned area

	November	December	January	February	March	April
Sugarcane burned area	58.79	35.12	64.48	46.69	64.04	13.66

■ Sugarcane burned area

Fig. 9 Monthly distribution of sugarcane burned area from Sentinel-2

Ranking No.	Sub-district	Sub-district area (km²)	Sugarcane plantation area (km²)	Sugarcane burned area (km²)	% Burned sugarcane over total sugarcane area
1	NONG SUANG	2125.08	9.07	3.50	39.96
2	NA KHAM	705.007	34.29	7.99	35.33
3	SAMRAN TAI	1702.38	21.19	7.04	34.93
4	BAYAO	1207.53	51.95	18.13	34.89
5	SAI THONG	1090.39	3.42	1.11	34.14
6	NONG KUNG THAP MA	439.073	29.55	8.37	33.59
7	HUAI MEK	989.013	19.87	7.64	33.53
8	PHIMUN	859.41	6.91	2.39	33.38
9	NONG NOK KHIAN	1441.41	19.23	6.56	33.09
10	PHU SING	764.102	3.61	1.23	32.84

Fig. 10 The top ten sub-districts with the highest percentage of sugarcane burned area over total sugarcane plantation area

3.5 Verification of Sentinel-2 Burned Area from Field Survey

In order to verify and confirm the ability of Sentinel-2's burned area detection in crop plots, a total of 16 burned plots from the field survey was collected at Kumphawapi and Nonsaat district in Udon Thani province. The verification area covered in Sentinel-2 tile ID T48QTD eight-couple images to create NBR from 14 December 2020 to 5 February 2021. The result presented that 15 of 16 verification plots could be detected by Sentinel-2. However, some of the burned periods from the field did not match the burned period of Sentinel-2 because the burned period from the field survey was only the assumed time.

Moreover, the burned and unburned plots were extracted from high-resolution imagery in Google Earth Pro to assess the Sentinel-2 burn areas. The Sentinel-2's burned images were selected to match with Google Earth's image date. There were 30 burned plots and 30 unburned plots that were used to validate the results. The validation result found that there was 24/30 detected burned pixel while 6/30 could not be detected. None of the verified unburned plots (0/30) were found as burned pixels. Therefore, the false negative was not found by Sentinel-2 burned area based on this study's current results.

3.6 Identify Sugarcane Harvesting Period and Harvest Methods

The identification estimated harvesting period of each sugarcane was analyzed by NDVI extracted from 15-day composite Sentinel-2 for 1 year in production season 2018/2019. Thus, Fig. 11 shows the number of harvested plots in the harvest season between the first half of November 2018 and the end of the following April. The total harvested sugarcane farms were 29,884 plots or 84% of the overall plots. From the beginning of harvesting season to the second half of December 2018, the number of harvesting plots fluctuated between 2163 and 2740 or around 55–94 km², and most sugarcane harvest activities were recorded in the first half of January 2019, a total of 8829 plots or approximately 224 km², while there were a small number of harvested plots in the second half of January, the first half of March, and the first half of April, with 731, 814, and 726 fields.

Moving on to the identification of harvest methods, the first half of April 2019 showed a small number of harvested plots, but most of them were harvested by burned method, accounting for 83% of the total harvested plots. At the same time, the first and second half of December 2018, the second half of January 2019, and the second half of February 2019 had a significant proportion of burned plots at just over three quarters, 65%, 75%, 73%, and 72%, respectively. On the other hand, the additional period had the majority proportion in cutting fresh sugarcane method, especially in the first half of November and the first half of March with approximately 80% of the total harvested plots.

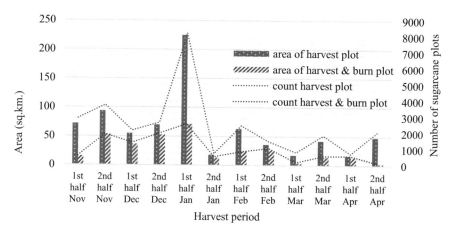

Fig. 11 The number of harvest plots and burn plots throughout the burning period

4   Discussion

In this study, the results of burned area derived from the finer resolution of Sentinel-2 presented more suitable than the coarse resolution of MODIS MCD64A1 in the small patch of the agricultural and field crop areas, since the size of agricultural fires in Southeast Asian countries was relatively small-scale, fragmented, and lasting a short duration. In contrast, MCD64A1 was better detected in the forest or the big cluster of burned areas, such as more extensive areas of rice residue burning to clear the land than sugarcane plots. Junpen also reported that the detection rate between the MCD64A1 and the Landsat-8 in the forest area was higher than cropland, accounting for 70% in forest and 41% cropland [13]. Thus, it is evident that most of the burned area in cropland from the reference fire grids did not overlap with the MCD64A1 product when compared to other types of land. Furthermore, the study found Landsat-8's could detect the more burned area in sugarcane plot around three times than MCD64A1 [14], compared to approximately 11 times higher detection between Sentinel-2's and MCD64A1 burned area in this study.

The verificiation with the field survey show that the dNBR index from Sentinel-2 could be precisely detected burned area in agricultural plot level from the validation assessment of the burned area. However, some validation plots could detect only a small part of burned area from the total plots. This might be because the damage of fires could not be detected by satellite sensors since the dNBR algorithms use land change detection from both imageries. Also, sugarcane plot boundaries that were requested from OCSB were found inconsistent with the actual plots. Figure 12

Fig. 12 Sugarcane boundary requested from OCSB compared to actual plots

illustrates the boundary of sugarcane plots showing in the yellow line provided by OCSB, which is composed of small fields inside when visualizing the actual area.

According to the harvested burned period of each plot, the result can estimate the sugarcane harvest period to fulfill the dataset and imply the harvested methods. Nonetheless, there was a big challenge in the temporal gap between harvested period and burned period from Sentinel-2. The problems were solved using day of the year (DOY) to reclassify as the same period, but some burned datasets had to be duplicated. The study reports that most sugarcane in the study area was harvested in the first half of January 2019 with cutting of fresh sugarcane, while in April 2019, there were a small number of harvested but most of the plots were found with burned harvest.

5 Conclusions

This study emphasized the acquiring of the burned area from MODIS MCD64A1 and Sentinel-2's NBR. The results showed that Normalized Burn Ratio (NBR) by Sentinel-2 was more suitable to detect the burned area in agricultural burned areas than MCD64A1 with almost 11 times higher detection rate. The total number of burned patched over the study period for MCD64A1 and Sentinel-2 is 221 km^2 and 2430 km^2. Based on the MCD64A1 product, there is an improvement in the ability to detect small fires in this version. Still, its resolution of 500-m per pixel was too coarse to detect fire in the agricultural area in this region with only big burned clusters from Sentinel-2 were detected, while burned areas derived from Sentinel-2 were more suited with the scale size of farm plot in Thailand.

It was found that most of the sugarcane in the study area was harvested in the first half of January, but most plots were harvested by cutting fresh sugarcane. The first half of April 2019 showed a small number of harvested plots, but most of them were harvested by burned method because of almost the end of the purchasing sugarcane period by sugar factories. The study also found that burn harvest was a major method instead of cutting fresh sugarcane in this area with over 70. Therefore, this study could fill the gap and emphasize the ability to present knowledge about crop residue burning over the region by using satellite images.

References

1. Webb, J., Hutchings, N., Amon, B.: Field burning of agricultural residues, pp. 1–14. European Environment Agency (EEA), Copenhagen (2013)
2. Crop, A., Tech, S., Dengia, A., Lantinga, E.: Advances in crop science and technology effect of pre-harvest cane burning on human health, soil quality and rate of cane weight loss in Ethiopian sugarcane plantations. Adv. Crop. Sci. Tech. **6**(5), 396 (2018)
3. Barrett, K., Kasischke, E.S.: Controls on variations in MODIS fire radiative power in Alaskan boreal forests: implications for fire severity conditions. Remote Sens. Environ. **130**, 171–181 (2013)

4. Junpen, A., Garivait, S., Bonnet, S.: Estimating emissions from forest fires in Thailand using MODIS active fire product and country specific data. Asia-Pac. J. Atmos. Sci. **49**(3), 389–400 (2013)
5. Portillo-Quintero, C., Sanchez-Azofeifa, A., Espirito-Santo, M.M.D.: Monitoring deforestation with MODIS Active Fires in Neotropical dry forests: an analysis of local-scale assessments in Mexico, Brazil, and Bolivia. J. Arid Environ. **97**, 150–159 (2013)
6. Venkatesh, K., Preethi, K., Ramesh, H.: Evaluating the effects of forest fire on water balance using fire susceptibility maps. Ecol. Indic. **110**, 105856 (2020)
7. Korontzi, S., Mccarty, J., Loboda, T., Kumar, S., Justice, C.: Global distribution of agricultural fires in croplands from 3 years of Moderate Resolution Imaging Spectroradiometer (MODIS) data. Glob. Biogeochem. Cycles. **20**(2), 1–15 (2006)
8. Ying, L., Shen, Z., Yang, M., Piao, S.: Wildfire detection probability of MODIS fire products under the constraint of environmental factors: a study based on confirmed ground wildfire records. Remote Sens. **11**(24), 3031 (2019)
9. Hantson, S., Padilla, M., Corti, D., Chuvieco, E.: Strengths and weaknesses of MODIS hotspots to characterize global fire occurrence. Remote Sens. Environ. **131**, 152–159 (2013)
10. Giglio, L., Boschetti, L., Roy, D.P., Humber, M.L., Justice, C.O.: The collection 6 MODIS burned area mapping algorithm and product. Remote Sens. Environ. **217**, 72–85 (2018)
11. Liu, T., Marlier, M.E., Karambelas, A., Jain, M., Singh, S., Singh, M.K., Gautam, R., DeFries, R.S.: Missing emissions from post-monsoon agricultural fires in northwestern India: regional limitations of MODIS burned area and active fire products. Environ. Res. Commun. **1**(1), 011007 (2019)
12. Kiadtikornthaweeyot, W., Sukawattanavijit, C., Rungsipanich, A.: Automatic detection of forest fire burnt scar from Landsat-8 image of northern part of Thailand. ECTI-CON 2018 – 15th International Conference on Electrical Engineering/Electronics, Computer, Telecommunications and Information Technology, 720–723 (2019)
13. Junpen, A., Pansuk, J., Garivait, S.: Estimation of reduced air emissions as a result of the implementation of the measure to reduce burned sugarcane in Thailand. Atmosphere. **11**(4), 366 (2020)
14. Pansuk, J., Junpen, A., Garivait, S.: Assessment of air pollution from household solid waste open burning in Thailand. Sustainability. **10**(7), 2553 (2018)

Machine Learning Approach with Environmental Pollution and Geospatial Information for Mapping Poverty in Thailand

Mahmud Isnan ⓘ, Teerayut Horanont ⓘ, and Anon Plangprasopchok ⓘ

Abstract Pollution is a major public health and human rights issue that dispropor-tionately affects the poor and disadvantaged. However, academic studies have not fully explored the relationship between poverty and the spatial distribution of environmental pollution, notably in Thailand. Thus, this chapter aims to fill a void in academic research by exploring the implementation of machine learning in the estimation of poverty by training input data from widely available and accessible open source, including environmental pollution and road density. The poverty rate is obtained from the TPMAP website and then clustered into two groups, "high and low," using hot spot analysis. The Google Earth Engine is used to extract pollution indexes such as carbon monoxide, formaldehyde, nitrogen, and sulfur dioxide, while road density is downloaded from OpenStreetMap. This study compares four different machine learning models: XGboost, lasso, random forest, and ridge regression for poverty estimation. The result of the study reveals that poverty-related areas are highly correlated with environmental pollution. The random forest technique has the best performance prediction among the four methods, with an R^2 of 0.79. Finally, feature importance analysis is used to determine the most influential features in order to assist decision-makers in gaining a better understanding of poverty.

Keywords Pollution · Poverty · Machine learning · Google Earth Engine · Geospatial regression

M. Isnan (✉) · T. Horanont
School of Information, Computer and Communication Technology, Sirindhorn International Institute of Technology, Thammasat University, Pathum Thani, Thailand
e-mail: m6322040871@g.siit.tu.ac.th; teerayut@siit.tu.ac.th

A. Plangprasopchok
National Electronics and Computer Technology Center, National Science and Technology Development Agency, Pathum Thani, Thailand
e-mail: Anon.Plangprasopchok@nectec.or.th

© The Author(s), under exclusive license to Springer Nature Switzerland AG 2023
W. Boonpook et al. (eds.), *Applied Geography and Geoinformatics for Sustainable Development*, Springer Geography,
https://doi.org/10.1007/978-3-031-16217-6_12

1   Introduction

Pollution is presently one of the world's most critical public health and human rights challenges, and it disproportionately affects the poor and disadvantaged. Pollution is not only an environmental issue; it also negatively influences the health and well-being of whole populations. Poverty and pollution are inextricably intertwined. Pollution is responsible for more than 92% of fatalities in low- and middle-income countries. Children are the most vulnerable victims of pollution because even small amounts of pollution can cause long-term health problems, premature death, decreased learning and earning capacity, and disability [1].

Pollution also has economic costs such as costs associated with pollution-related diseases, including direct medical fees, costs to the healthcare system, and the potential costs associated with missed productivity and economic development. The estimated annual cost of pollution-related welfare losses is $4.6 trillion, or 6.2% of global economic production. It is one of the top causes of death globally, and in low- and middle-income countries, it is responsible for 92% of all deaths. Polluted air, soil, and water are frequently a relentless toxic attack on people in the poorest, most vulnerable areas. Achieving the global goals, also known as the Sustainable Development Goals (SDGs), will require concerted efforts to combat pollution because of the many ways that pollution impacts the health and livelihoods of people around the world. According to the Institute for Health Metrics Evaluation, pollution-related mortality accounts for 9 million deaths annually. That is the equivalent of all the people in Manhattan, Bangkok, or Bogota dying year after year. Millions more, particularly children, are sickened or disabled due to harmful environmental exposure [2]. Catherine Ganzleben, director of the European Environment Agency's air pollution and environmental health sections, spoke with the Borgen Project (EEA). "Pollution affects disadvantaged populations more than wealthier ones due to a lack of access to medical treatment and exposure to climate change effects," she added. Even when wealthy and underprivileged persons are exposed to the same amount of pollution, the needy are more affected. Lower-income residents in polluted areas are shown to be more vulnerable to the health consequences of pollution than wealthier residents of polluted areas [3].

The use of machine learning for poverty estimation has been proven in several previous studies. For instance, researchers from the Asian Development Bank (ADB) improved the conventional small area poverty estimation framework by incorporating geospatial data from daytime and nighttime imagery through machine learning algorithms to produce granular poverty maps rather than the traditional poverty estimates [4–6]. In further research published in 2020, the association between survey-based socioeconomic circumstances and the value of the nighttime light (NTL) index was statistically validated, and the dynamics of regional inequality were explored using both NTL and survey-based indices [7]. Furthermore, research has identified correlations in the regional variation of poverty intensity and normalized difference vegetation index (NDVI) from a spatial standpoint; poverty intensity (and hence nutrition and child mortality) changes inversely with NDVI [8].

Using ordinary least squares (OLS) and geographically weighted regression (GWR) models, another research discovered that poverty-related locations were significantly associated with positive high and/or negative high vegetation in both decreasing and increasing cities [9]. Similar models also were applied in Sichuan province and India to investigate the relationship between poverty and physical geographic factor [10, 11]. Morikawa's findings revealed a substantial correlation between the existence of saving organizations and a favorable change in vegetation [12].

Poverty drives pollution, and pollution drives poverty. One of the major ways in which poverty contributes to environmental damage is deforestation. Forests provide clean air to the Earth while also functioning as "sinkholes" to help buffer the world's present catastrophic climate upheavals. The rural poor are generally situated in vulnerable ecosystems and must rely heavily on natural resources such as forest wood to live daily. This causes further deforestation, making it more difficult for them to escape the poverty trap [13–15]. Furthermore, the poor are more vulnerable to health risks because their economic condition may limit access to health treatment when health problems result from pollution exposure, resulting in low productivity at work. However, academic research has not thoroughly studied the relationship between poverty and the geographical distribution of environmental pollution, notably in Thailand. Thus, this chapter aims to contribute to filling a void in academic research by examining poverty differences and inequities associated with environmental pollution at the sub-district level for 2019. This will be accomplished through the use of machine learning and the integration of pollution data derived from remote sensing with socioeconomic data. This chapter is to utilize machine learning models such as XGboost, lasso, random forest, and ridge regression.

2   Datasets

2.1   Poverty Data

The poverty rate at the sub-district level in 2019 is obtained from the Thai People Map and Analytics Platform (TPMAP) website. TPMAP is a data analytics platform dedicated to the precise alleviation of poverty in Thailand [16].

2.2   Environmental Pollution

We utilize Google Earth Engine (GEE) to extract pollution index from publicly available real-time high-resolution satellite pictures and use them as features, including the following: (1) Carbon monoxide (CO) is an important atmospheric trace gas. It is a significant pollutant in several cities. CO is produced by burning fossil fuels, biomass, methane, and other hydrocarbons in the atmosphere [17]. (2) Formaldehyde is an intermediate gas in all non-methane volatile organic compound (NMVOC) oxidation cycles, leading to CO_2 [18]. (3) Sulfur dioxide, sometimes known as SO_2, is a chemical compound with SO's formula. It is the poisonous gas responsible for the smell of burning matches. It is produced as a by-product of copper extraction and the burning of sulfur-containing fossil fuels and is released naturally by volcanic activity [19]. (4) Nitrogen oxides (NO_2 and NO) are major trace gases in the troposphere and stratosphere. Anthropogenic (especially fossil fuel and biomass combustion) and natural processes release them into the atmosphere (wildfires, lightning, and microbiological processes in soils) [20]. Data are taken from Sentinel-5P by the European Space Agency. We calculate the mean, max, min, median, variance, skew, kurtosis, and standard deviation for each satellite image.

2.3   Road Density

We acquire crowd-sourced geospatial data, i.e., road density, from OpenStreetMaps (OSM) via the Geofabrik online repository since vehicle on the road is a potential source of air pollution [21]. Then, we use Zonal Statistics as table tool in ArcGIS to extract the mean road density at the sub-district level in Thailand.

3   Proposed Methodology

The overall methodological framework of the study to solve the problem is illustrated in Fig. 1. As alternative to costly high-resolution satellite imagery, this study focuses on the application of open-source data as shown in Table 1. We use the SHAPE file of sub-district level to extract input features, which are independent

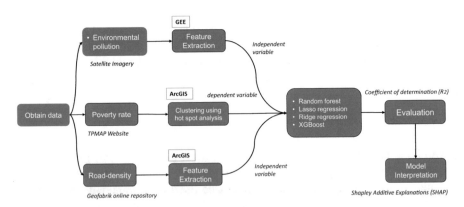

Fig. 1 Overall flowchart of methodological framework

Table 1 Dependent and independent variables derived from multi-source data

Variable	Descriptions	Source
	Dependent	
Poverty rate	Poors in the corresponding tumbol (sub-district)	TPMAP website
	Independent	
Carbon monoxide (CO)	The mean, max, min, median, variance, skew, kurtosis, and standard deviation of carbon monoxide in each sub-district	Google earth engine
Formaldehyde	The mean, max, min, median, variance, skew, kurtosis, and standard deviation of formaldehyde in each sub-district	Google earth engine
Sulfur dioxide	The mean, max, min, median, variance, skew, kurtosis, and standard deviation of sulfur dioxide in each sub-district	Google earth engine
Nitrogen	The mean, max, min, median, variance, skew, kurtosis, and standard deviation of nitrogen in each sub-district	Google earth engine
Road density	The total length of primary and secondary roads in each sub-district	OpenStreetMaps

variables, from environmental pollution and road density to estimate poverty. The original poverty data is supplied by the TPMAP website, which was transformed into two clusters of "high and low" poverty throughout the process of clustering using hot spot analysis in ArcGIS software to avoid overfitting during the training model. Hot spot analysis is a geographical analysis and mapping approach that looks for spatial grouping. Points on a map represent locations of events or objects [22]. The Getis-Ord Gi* statistic (pronounced G-i-star) is calculated for each feature in a dataset. The resulting z-scores and p-values show where high and low values cluster geographically. The mean score is then used as a dependent variable. Hot spot analysis can be seen in Fig. 2.

XGboost, lasso, random forest, and ridge regression are the four main categories of machine learning models in this approach. We chose the coefficient of determination (R^2) to assess the models' performance which provides the goodness of fit of a set prediction to the actual values. The value of R^2 is between 0 and 1 for no-fit and perfect fit on the regression line [23]. Then, we interpret the model using SHAPley Additive Explanations (SHAP). Finally, the predicted poverty map is generated using the model with the highest performance.

4   Experiment and Result

In order to ensure that features of the system make sense roughly as assumed, we perform a sanity check on all extracted features before training them into machine learning model using the Pearson correlation coefficient. In general, environmental pollution indexes are negatively associated with the poverty rate, apart from the

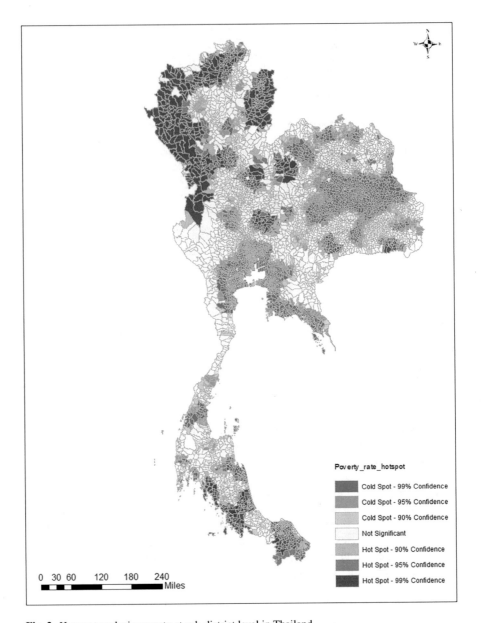

Fig. 2 Hot spot analysis poverty at sub-district level in Thailand

mean and standard deviation of carbon monoxide as well as the mean and median sulfur. On the other hand, road density is positively correlated with poverty. We then standardize all input features to improve the efficiency of the models by scaling numerical input variables to a standard range using the popular scikit-learn library in Python. We split the data into 80:20 ratios for training and testing, respectively. A tenfold cross-validation is used to evaluate all models. We fine-tune and optimize the parameters and hyperparameters by cross-validated search over parameter settings (randomizedsearchcv), as shown in Table 2, to achieve the best performance.

In Fig. 3, we compare the R^2 values for the various models discussed in section "Proposed methodology" and discover that the random forest regression model

Table 2 The descriptions and values of random forest regression (RFR) model parameters

Parameter names	Descriptions	Values
n_estimators	The number of trees in random forest	70
max_depth	The maximum depth of tree	80
min_sample_split	The minimum number of samples required to split an internal node	5
min_sample_leaft	The minimum number of samples required to be at a leaf node	2
Boostrap	Method for sampling data points (with or without replacement)	False

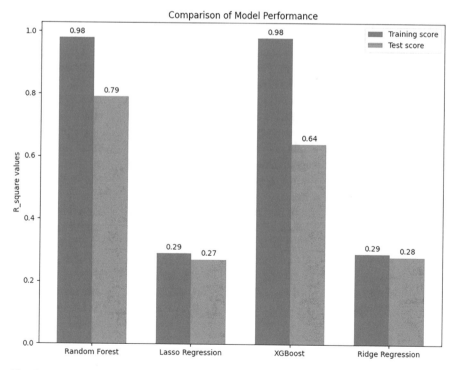

Fig. 3 Model performance (R^2) from dependent variable (poverty) and independent variables (environmental pollution and road density)

outperforms them all, with an R^2 of 0.79. We also compute only NTL derived from GEE and clustered poverty at the sub-district level to see whether our model is accurate as previous studies using random forest and find the R^2 is 0.66. We note our result for poverty estimation marginally exceeds past performance benchmarks for the same task conducted in other Southeast Asian countries, 0.57 for Myanmar [23] and 0.63 for the Philippines [24]. However, the previous study conducted in Thailand by Puttanapong et al. utilizing a combination of NTL, NDVI, land surface temperature, and OSM data such as point of interest, road length, road count, and built-up area reveal R^2 is 0.83 [5]. This finding indicates that a combination of environmental pollution has better performance than using only the NTL index, while utilizing some combination with NTL will improve the performance.

5 Discussion

This study explores the application of open-source data and machine learning approach for poverty estimation using features directly extracted from Google Earth Engine. Our model achieves the encouraging result of R^2, 0.79 in poverty prediction. The result generally agrees with domain knowledge of the on-the-ground situation in Thailand. The lower pollutions indicate a higher poverty rate because pollution is usually produced by vehicles such as motorbikes and cars, which are very expensive. As proof of the concept, we show in Fig. 4 a predicted poverty map of the sub-district level using random forest. Red "hot" areas represent statistically significant poverty compared to blue "cold" areas with low poverty.

We quantify the influence of each attribute on a single prediction using SHAPley Additive Explanations (SHAP). As an example, we depict in Fig. 5 the shape values for anticipated poverty of one sub-district in Thailand. Here, the minimum and median carbon monoxide values increase the predicted poverty. On the other hand, the mean and median nitrogen values lead to a decrease in predicted poverty.

We compute SHAP feature importance, as shown in Fig. 6. It shows that maximum carbon monoxide is the most important variable, with an important weight of 0.39. Among the four types of environmental pollutants, carbon monoxide (including max, min, variance, median, mean, and std) is the most important variable with a total important weight of 1.0, indicating that carbon monoxide is the key factor in estimating poverty rate. On the other hand, nitrogen (including mean, max, median, min, variance, and std) is the second position as the most important variable, with a collective importance weight of 0.88. In contrast, sulfur dioxide is the least contributing to estimating poverty, with a marginal importance weight of 0.13.

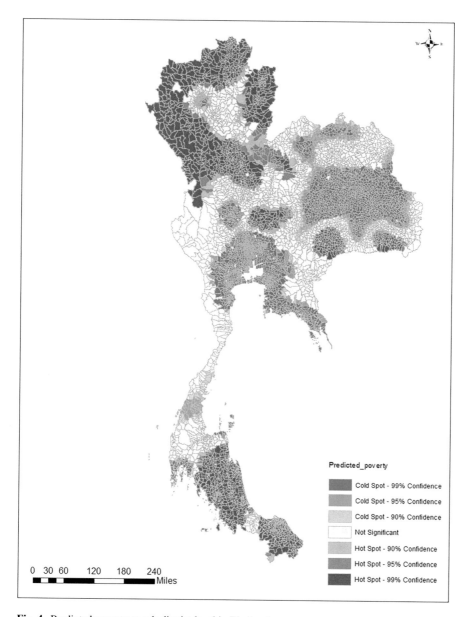

Fig. 4 Predicted poverty at sub-district level in Thailand

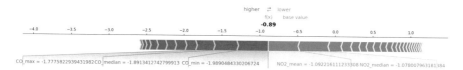

Fig. 5 Example of SHAP values for predicted poverty of one sub-district level. The color represents whether a variable increases the predicted value (red) or decreases the value (blue)

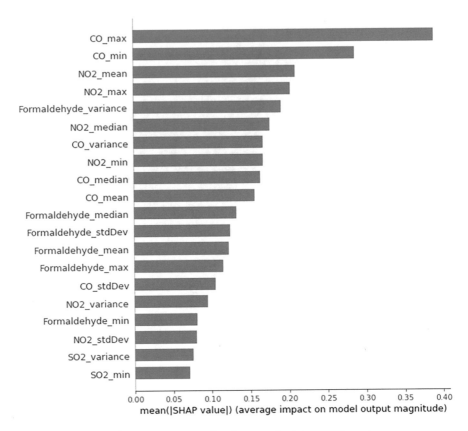

Fig. 6 SHAP feature importance measured as the mean absolute SHAPley values

6 Conclusion

In this chapter, we have proposed an alternative approach to poverty estimation using cost-efficient machine learning methods combined with environmental pollution extracted from Google Earth Engine and road density at the sub-district level. In data preparation, we cluster the poverty rate into two groups, "high and low" poverty, using hot spot analysis in ArcGIS software to avoid overfitting. We compare four different machine learning and determine that random forest is the most accurate prediction method in the case of Thailand due to its capability to handle multicollinearity problems, yielding an accuracy of 79%. We also compute the model with NTL data only and find our model outperforms the same task conducted in other Southeast Asian countries. This contribution suggests the potential of applying the environmental pollution extracted from satellite imagery to analyze the spatial distribution of poverty since they are highly correlated. Moving forward, future studies are required to investigate whether such a relationship can be shown to be constant over time, so it might be conceivable to apply this method to forecast poverty during years when household income and expenditure surveys are not completed, but geospatial data correlates poverty are available. Finally, using SHAP to assess model interpretability, we determine the most predictive global indicator of poverty to max carbon monoxide. Interpretable poverty maps are vital for policy-makers and development organizations to understand the geographical distribution of poverty better and to implement targeted humanitarian programs and initiatives to alleviate it where it exists.

Acknowledgments This research is financially supported by Thailand Advanced Institute of Science and Technology (TAIST), National Science and Technology Development Agency (NSTDA), Tokyo Institute of Technology, and Sirindhorn International Institute of Technology (SIIT), Thammasat University (TU) under the TAIST-Tokyo Tech Program. Also this research is partially supported by Thammasat University Research fund under the TSRI, Contract No. TUFF19/2564 and TUFF24/2565, for the project of "AI Ready City Networking in RUN," based on the RUN Digital Cluster collaboration scheme.

References

1. Heart (2017) Pollution and Poverty – HEART. https://www.heart-resources.org/reading_pack/pollution-and-poverty/. Accessed 18 Nov 2021
2. GAHP (2017) Achieving The #GlobalGoals And Addressing Pollution Go Hand In Hand – GAHP | Global Alliance on Health and Pollution. https://gahp.net/achieving-globalgoals-addressing-pollution-go-hand-hand/. Accessed 9 May 2022
3. Borgen Project (2021) How Air Pollution Affects Poverty in Europe – The Borgen Project. https://borgenproject.org/air-pollution-and-poverty/. Accessed 18 Nov 2021
4. ADB: Mapping Poverty Through Data Integration and Artificial Intelligence (2020)
5. Puttanapong, N., Martinez Jr, A., Addawe, M., et al.: Predicting Poverty Using Geospatial Data in Thailand working paper series (2020)
6. ADB: Mapping the Spatial Distribution of Poverty Using Satellite Imagery in Thailand, (2021)
7. Sangkasem, K.: Analysis of spatial inequality using DMSP-OLS nighttime-light satellite imageries: a case study of Thailand. Wiley Online Libr. (2020). https://doi.org/10.1111/rsp3.12386
8. Sedda, L., Tatema, A.J., Morley, D.W., et al.: Poverty, health and satellite-derived vegetation indices: their inter-spatial relationship in West Africa. Int. Health. **7**, 99–106 (2015). https://doi.org/10.1093/inthealth/ihv005
9. Dawson, T., Onésimo Sandoval, J.S., Sagan, V., Crawford, T.: A spatial analysis of the relationship between vegetation and poverty. ISPRS Int J Geo-Information. **7** (2018). https://doi.org/10.3390/ijgi7030083
10. He, X., Mai, X.: Poverty and physical geographic factors: An empirical analysis of Sichuan Province using the GWR model. Sustainability. (2020). https://doi.org/10.3390/su1301
11. Bhattacharya, H., Innes, R.: Is there a Nexus between Poverty and Environment in Rural India? 2006 Annu Meet July 23–26, Long Beach (2006)
12. Morikawa, R.: Remote sensing tools for evaluating poverty alleviation projects: a case study in Tanzania. Procedia Eng. **78**, 178–187 (2014). https://doi.org/10.1016/j.proeng.2014.07.055
13. Chen, S., Ou, J.: The environmental and health impacts of poverty alleviation in China: From a consumption-based perspective. Sustainability. (2021). https://doi.org/10.3390/su13041784
14. Casillas, C.: The energy-poverty-climate nexus. Science. **330**, 1181–1182 (2010). https://doi.org/10.1126/science.1197412
15. Dasgupta, S., Laplante, B., Wang, H., Wheeler, D.: Confronting the environmental Kuznets curve. J. Econ. Perspect. **16**, 147–168 (2002). https://doi.org/10.1257/0895330027157

16. Dorji, U., Siripanpornchana, C., Surasvadi, N., et al.: Exploring night light as proxy for poverty and income inequality approximation in Thailand. In: IEEE region 10 annual international conference, proceedings/TENCON, pp. 1082–1087 (2019)
17. Bakwin, P.S., Tans, P.P., Novelli, P.C.: Carbon monoxide budget in the northern hemisphere. Geophys. Res. Lett. **21**, 433–436 (1994). https://doi.org/10.1029/94GL00006
18. Sinha, V., Hakkim, H., Kumar, V.: Advances in identification and quantification of non-methane volatile organic compounds emitted from biomass fires through laboratory fire experiments. Adv Atmos Chem, 169–197 (2019). https://doi.org/10.1142/9789813271838_0003
19. Craig, K.: A review of the chemistry, pesticide use, and environmental fate of sulfur dioxide, as used in California. Rev. Environ. Contam. Toxicol. **246**, 33–64 (2019). https://doi.org/10.1007/398_2018_11
20. Thompson, A.M.: The oxidizing capacity of the Earth's atmosphere: probable past and future changes. Science. **256**, 1157–1165 (1992). https://doi.org/10.1126/SCIENCE.256.5060.1157
21. Xue, J., Liu, S., Chen F.: GIS analysis on relationship between traffic density parameters and near-road air pollutants (2019)
22. Rachiel, T.: Hot spot spatial analysis Columbia public health. In: Columbia Public Heal (2020) https://www.publichealth.columbia.edu/research/population-health-methods/hot-spot-spatial-analysis. Accessed 26 May 2022
23. Lin Htet, N., Kongprawechnon, W., Thajchayapong, S., Isshiki, T.: Machine learning approach with multiple open-source data for mapping and prediction of poverty in Myanmar. In: ECTI-CON 2021–2021 18th International Conference on Electrical Engineering/Electronics, Computer, Telecommunications and Information Technology: Smart Electrical System and Technology, Proceedings, pp. 1041–1045 (2021)
24. Tingzon, I., Orden, A., Go, K.T., et al.: Mapping poverty in the philippines using machine learning, satellite imagery, and crowd-sourced geospatial information. Int Arch Photogramm Remote Sens Spat Inf Sci ISPRS Arch. **42**, 425–431 (2019). https://doi.org/10.5194/isprs-archives-XLII-4-W19-425-2019

Integration of Machine Learning Algorithms and Time-Series Satellite Images on Land Use/Land Cover Mapping with Google Earth Engine

Guntaga Logavitool ⓘ, Kritchayan Intarat ⓘ, and Teerayut Horanont ⓘ

Abstract This study demonstrates the usage of the Google Earth Engine (GEE) cloud service for LULC classification in Nakhon Nayok, Thailand. Herein, multi-temporal Sentinel-2 images are incorporated with supervised machine learning algorithms in order to determine seven land use and land cover (LULC) classes. Pixel-based (PB) and object-based (OB) classification methods are also considered and evaluated. When using the PB approach combined with the random forest algorithms and median composited time-series data, results demonstrated that the highest attainment is achieved: namely, overall accuracy of 85.42% and kappa statistics reaches 0.82. Z-statistics confirm the potential of median composited datasets that provide significant performance (p-value <0.05). GEE reveals many advantages in geospatial analysis from this experiment. GEE is a robust cloud-serviced tool for LULC time-series image classification.

Keywords Remote sensing · LULC · Machine learning · Time-series analysis · Google Earth Engine

G. Logavitool
School of Information, Computer and Communication Technology, Sirindhorn International Institute of Technology, Pathum Thani, Thailand

Faculty of Liberal Arts, Department of Geography, Thammasat University, Pathum Thani, Thailand

K. Intarat (✉)
Faculty of Liberal Arts, Department of Geography, Thammasat University, Pathum Thani, Thailand
e-mail: intaratt@tu.ac.th

T. Horanont
School of Information, Computer and Communication Technology, Sirindhorn International Institute of Technology, Pathum Thani, Thailand

© The Author(s), under exclusive license to Springer Nature Switzerland AG 2023
W. Boonpook et al. (eds.), *Applied Geography and Geoinformatics for Sustainable Development*, Springer Geography,
https://doi.org/10.1007/978-3-031-16217-6_13

1   Introduction

Socioeconomic and natural activities play significant roles in the variation of the earth's surface [1, 2]. Remote sensing is a primary source and extensively associated with land use mapping [3]. Remote sensing is recognized as the basis of further analysis, such as land monitoring and area management. Acquiring an up-to-date LULC map is necessary for effective local administration. Due to climate variability and crop diversity, agricultural activities tend to decrease causing LULC to change rapidly; acquiring a precise and accurate LULC map is challenging. Remote sensing has been introduced in association with LULC mapping [3, 4]. The capability of massive area coverage and multispectral band combination of remotely sensed data can sufficiently support LULC classification.

LULC change can be determined by analyzing satellite images and can be done via remote sensing. Remote sensing helps to understand environmental change dynamics in order to ensure sustainable development. Different spectral reflectance from each period can potentially discriminate the characteristics of LULC variation [5]. A series of multi-satellite images can be adopted to categorize LULC. Remote sensing requires considerable resources and special sensors. Vast storage space for images is needed. Computing units must have sufficient capacity to deal with image processing. Dealing with high-dimensional analysis requires high-performance software. Each image can contain a huge amount of data: hundreds of megabytes to gigabytes. It is acknowledged that acquiring images through the providers' portals takes much effort and time [6–8].

Recently, cutting-edge cloud computing technology has been developed and utilized in satellite image processing. Since 2010, GEE cloud mapping has been a preference (https://earthengine.google.com). GEE provides a simplistic way to compute and evaluate geospatial data. For example, planetary-scale data of publicly available satellite imagery and scientific datasets is freely used and regularly updated [9–11]. Users will experience diverse data and end-to-end analytical solutions through JavaScript, Python, or R programming languages, which do not require advanced skills in remote sensing analysis software, such as ENVI, ERDAS, and eCognition [9].

Machine learning algorithms, known as non-parametric algorithms, have also been applied to classify remotely sensed data [12] such as artificial neural networks (ANN), decision tree (DT), random forest (RF), as well as support vector machine (SVM). For image classification, the pixel-based approach is commonly used and relies on spectral information for containing individual pixels. The recent development of image classification techniques makes the object-based (OB) approach much more popular than traditional techniques [11]. The OB approach counts on the group of pixels' characteristics regarding similarity; classification is carried out per object rather than pixel [13]. Previous studies [14–16] suggest that the OB approach provides better results as regards higher-resolution data. It is noted that the PB approach is recommended for lower resolution since the shape of an object is quite large [13], such as agricultural areas.

This study aims to evaluate the performance of LULC classification techniques using GEE by integrating the non-parametric machine learning classifiers (CART and RF) along with multi-temporal satellite images. Analytical OB and PB supervised methods are applied to discriminate each land use category in Nakhon Nayok, Thailand.

2 Materials and Methods

2.1 Study Area

Nakhon Nayok Province lies in the central region of Thailand, at 14° 12′ 44.95″ N, 101° 12′ 06.57″ E (Fig. 1). It is approximately a 1.5-h drive from Bangkok and covers 2122 sq. km. The topography of the central and southern part exhibits a flood plain terrain, which is mainly agricultural covering 55.70% of the total area. Rice is the main crop harvested taking up 38.86% of Nakhon Nayok Province. The province's northern area reveals a mountainous area where Khao Yai National Park is located, enveloping 30.26% of the area. According to Koppen's classification, the climatology of this location is defined as tropical savanna climate (Aw); average temperature is 23.8 °C, and precipitation is around 1916 mm/year. LULC

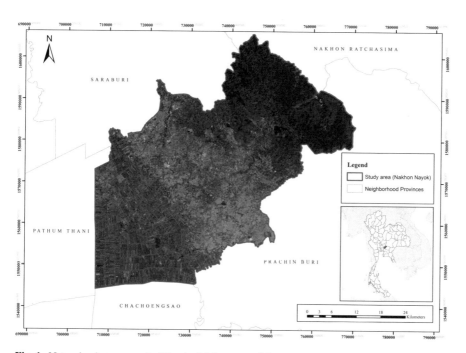

Fig. 1 Natural color-composited Sentinel-2 imagery of the study area

classification in this study comprises seven classes: paddy rice, perennial tree, orchard, forest, bare land, built-up area, and water body.

2.2 Data Acquisition and Preprocessing

Analytical data is comprised of (1) multi-temporal satellite data, (2) ancillary data, and (3) field observation. This study used the Sentinel-2 satellite imagery provided by the European Space Agency (ESA) from January to December 2020 to construct the time-series data. The band combination consisted of four spectral bands, e.g. blue, green, red-edge 2, and near-infrared, which included vegetation indices. Normalized difference vegetation and normalized difference red-edge indices (NDVI and NDRE) were implemented as additional enhanced layers, as expressed in Eqs. 1 and 2:

$$NDVI = \frac{(NIR - RED)}{(NIR + RED)} \tag{1}$$

$$NDRE = \frac{(NIR - RE2)}{(NIR + RE2)} \tag{2}$$

where NIR, RED, and RE2 represent the reflection of near-infrared, red, and red-edge 2 spectrums.

The preferred level 2 products were atmospherically corrected from Top-of-Atmosphere (TOA) to Bottom-of-Atmosphere (BOA) for the entire pixel [17]. Sentinel-2 delivered the QA60 band to improve the cloud issue by masking all cloud coverage and shadowed areas; cloud cover area should be less than 20%. A practical tool can be obtained by choosing the s2cloudless built-in function.

Ancillary data, such as the Nakhon Nayok administration boundary and base LULC map, were acquired from the Land Development Department (LDD). Both training and validation samples, i.e., 585 and 141, respectively, were randomly selected from the LDD-based LULC. Subsequently, the training dataset applied a ratio of 80/20 in order to construct the testing dataset. In February 2021, a field survey was carried out to inspect the sampling data and identify the ground validation (Fig. 2).

2.3 Dataset Composition

Time-series satellite data can improve LULC change detection and determine the cloud cover issue regularly presented in tropical regions [18]. It is seen that Sentinel-2 delivered 5 days of revisited imagery by combining Sentinel-2A and 2B

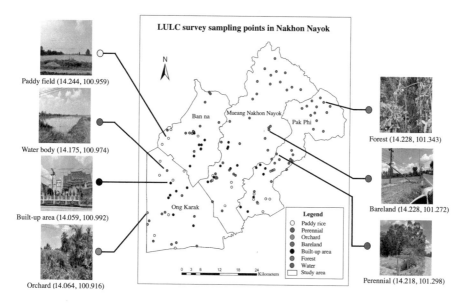

Fig. 2 LULC survey sampling points in Nakhon Nayok, Thailand

constellations, thus contributing six images each month at the exact location. The benefit of such frequent acquisitions ensured that missing data can be fulfilled [14].

This experiment applied the time-series composition, focusing on central tendency statistics and non-parametric classifiers to discriminate the LULC class. The investigation determined the mean values of multiple layers, e.g., blue, green, RE2, NIR, NDVI, and NDRE along with the standard deviations of NDVI and NDRE. Datasets were then statistically associated with the mean and median reduced region operation (D1 and D2) and masked by the province's administration.

2.4 Non-parametric Machine Learning Classifiers

The classifiers' selection depended on many options: type of data, statistical distribution of classes, target accuracy, ease of use, speed, scalability, and interpretability [12]. For the classification, non-parametric machine learning classifiers were proposed, which resulted in greater flexibility and proficiency by fitting in a significant number of functional applications. Thus, ML algorithms were able to freely learn the functional pattern from the training dataset. Due to their remarkable abilities, the algorithms provided a higher performance when executing the predictions [19, 20]. In this analysis, well-known machine learning classifiers were selected in accordance with GEE (ee.Classifier). Classification and regression tree (CART) and random forest (RF) classifiers were thus employed and used to evaluate results when working with both PB and OB classifications.

It is noted that CART, an advanced technique of decision trees, consisted of regression and classification trees. Such an algorithm has the advantage of dealing with various types of data, generating a complex decision flowchart similar to a tree structure through learning the input dataset and can be expressed as follows:

$$G(x) = \sum_{t=1}^{T} q_t(x) \cdot g_t(x) \tag{3}$$

where $g_t(x)$ refers to the base hypothesis, which represents the leaf at the end of the path t, and $q_t(x)$ is a condition for determining the appropriate path (t) of x.

As for uncertain decision trees and noise sensitivity leading to overfitting, CART can prove to be disadvantageous [21–23]. Thus, CART parameters were examined for acquiring the highest accuracy before classification. According to the discrepancy found in CART, RF was put forward in order to overcome the issue of overfitting, combining many trees and employed a probabilistic scheme by majority vote as revealed below:

$$RFfi_i = \frac{\sum_j normfi_{ij}}{\sum_{j \in all\ features, k \in all\ trees} normfi_{jk}} \tag{4}$$

where $RFfi_i$ is the importance of feature i determined from all trees in the model and $normfi_{ij}$ represents the normalized feature importance for i in tree j.

Despite unknown decision rules, RF is seen to reduce overfitting and estimation of biased accuracy [24–26]. Hence, parameters such as the number of trees, minimum leaf population, and the maximum number of leaf nodes are seen to reach the best classification result.

2.5 Classification and Accuracy Assessment

Consequently, results of both classifications were evaluated using confusion matrices. Overall accuracies (OA) were computed for all classifications and used to estimate the performance of both CART and RF. Cohen's Kappa coefficient (K) was also considered for judging the reliability of the results [27]. In order to justify the performance between the competing classifiers, relative comparisons were conducted by taking into consideration the Z-scores [28]. In Fig. 3, the workflow of the experiment is shown.

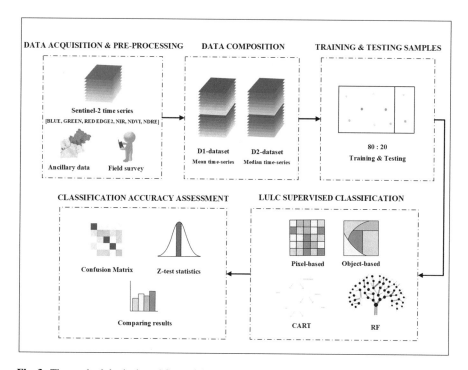

Fig. 3 The methodological workflow of the classification

Table 1 Parameters and value settings for ML model evaluation

Parameters	Values
Maximum number of leaf nodes (P1)	1, 2, 3, 4, 5, 6, 7, 8, 9, 10, 20, 30, 100, 500, 1000
Minimum leaf population (P2)	1, 2, 3, 4, 5, 6, 7, 8, 9, 10, 20, 30, 100, 500, 1000
Number of trees (P3)	10, 20, 50, 100, 200, 300, 400, 500, 600, 700, 800, 900, 1000

3   Results and Discussion

3.1   Parameters of the Machine Learning Classifiers

The ML parameters of CART and RF models were considered to be the best combination of parameters. Moreover, each classifier model was evaluated both by PB
and OB via the statistical reducers (mean and median). Parameters determined in
this section involved the maximum number of leaf nodes and the minimum leaf
population. The RF classifier included an additional parameter (the number of trees)
for significant majority voting. In Table 1, the testing values for each parameter are
presented. In Figs. 4 and 5, the results using tenfold cross-validation are shown.

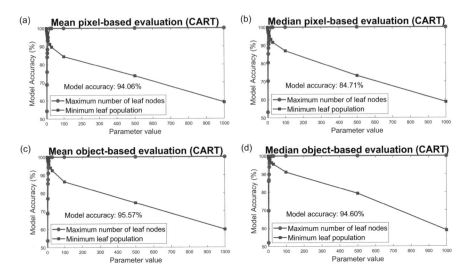

Fig. 4 CART model optimization for classification methods. PB mean and median reducers are displayed in (**a**) and (**b**). OB mean and median reducers are displayed in (**c**) and (**d**)

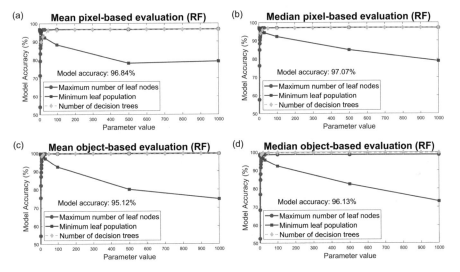

Fig. 5 RF model optimization for classification methods. PB mean and median reducers are displayed in (**a**) and (**b**). OB mean and median reducers are displayed in (**c**) and (**d**)

According to Fig. 4, OB classifications delivered higher CART model accuracies. The mean reducer technique provided the highest performance attaining a testing accuracy of 95.57%. The best parameter combinations, i.e., P1 and P2, proved to be 9 and 1, respectively. The median reducer demonstrated a comparable value of 94.60%. However, the PB classifications of the CART model were less efficient than those of OB. Both mean and median reducers were reported to be 94.06% and 84.71%, respectively.

In Fig. 5, the RF model parameter testing included P3 (number of trees) for consideration. Hence, the highest model performance was achieved delivering 97.04% accuracy via the median reducer. PB classification using the mean reducer reached 96.84%, a high achievement similar to the median reducer.

3.2   Classification Results and Accuracies

In Table 2, eight validation results derived from the confusion matrices are reported.

With reference to Table 2, the PB method in conjunction with the RF classifier outperformed other classifications. It is seen that the PB method produced the best OA when working on the D2 dataset reaching 85.42%. The K measurement (0.82) further indicated a high degree of consistency. When considering RF with OB classification, less accuracy was achieved. Furthermore, the CART algorithm provided an excellent classification result, viz., 80.56% (OA). Besides, the Kappa statistic (0.76) indicated considerable consistency.

CART also performed well with D2 and OB classification, achieving 77.08% OA and 0.72 K coefficient. In Fig. 6, all experiments present different classified maps that visually describe the areas of study.

The statistical Z-score (Fig. 3), denoting the highest classification result (PB_ D2_RF), demonstrated excellent performance when compared to three out of 4 results acquired by the D1 dataset (PB_D1_CART, OB_D1_CART, and OB_D1_ RF). Such an outcome indicates that both PB and RF perform well with the composited time-series data (D1 and D2). However, the null hypothesis cannot be rejected compared to the PB_D1_RF and all combinations that used the D2 dataset.

In general, the OB approach prefers to incorporate a very-high-resolution image, namely, medium-to-high-resolution image [15]. In addition, OB requires more extensive computational resources and complex parameter settings such as size, segmentation, and compactness, while PB requires only spectral information to analyze and classify [29, 30]. In this study, the PB approach reveals greater classification performance than OB.

Table 2 OA and K as determined via eight experimental classifications

Dataset	Classifier	PB		OB	
		OA (%)	K	OA (%)	K
D1 (mean)	CART	74.30	0.69	76.50	0.54
	RF	78.47	0.74	76.39	0.71
D2 (median)	CART	80.56	0.76	77.08	0.72
	RF	85.42	0.82	78.47	0.78

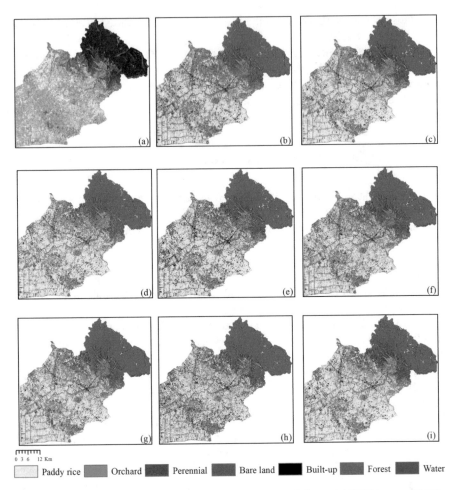

0 3 6   12 Km

☐ Paddy rice   ▉ Orchard   ▉ Perennial   ▉ Bare land   ▉ Built-up   ▉ Forest   ▉ Water

Fig. 6 The classified maps from the eight experiments are as follows. (**a**) RGB image, (**b**) PB_D1_CART, (**c**) PB_D1_RF, (**d**) OB_D1_CART, (**e**) OB_D1_RF, (**f**) PB_D2_CART, (**g**) PB_D2_RF, (**h**) OB_D2_CART, and (**i**) OB_D2_RF

3.3 Performance of the GEE Platform: Time-Series Data Analysis

In the remotely sensed time-series data analysis, the GEE cloud-computing platform revealed high efficiency. Such a platform allowed users to employ computing resources and extensive data archives covering planetary-scale earth science data and analysis [21]. Especially, satellite time-series data require a great deal of storage capacity, high downloading speed, and high computing resources. It is noted that GEE overcame these issues [31]. However, parameter testing is still unavailable via GEE and needs to be handled by other programming languages. GEE's time-series composited function (ee.Reducer) only incorporates standard statistics, e.g., maximum, minimum, mean, median, mode, and percentile to aggregate the frequency and dimension of data. When using GEE, some limitations need to be considered.

4 Conclusion

In this paper, we evaluate the performance of LULC classification techniques using GEE by integrating the non-parametric machine learning classifiers (CART and RF) and multi-temporal satellite images. PB and OB approaches were selected and implemented applying two ML classification algorithms, RF and CART, together with two composited data (D1 and D2). Consequently, eight classifications were carried out, resulting in eight classified maps. Accuracy assessments from the results were found, using confusion matrix and Z-statistical testing. The highest accuracy was determined using the PB approach combined with the RF algorithm and D2. The GEE platform is seen to allow all users access to archived data and provides cloud resources for geospatial operation. It is evident that GEE can be applied as a potent open cloud-sourcing tool toward conventional time-series LULC classification.

References

1. Jensen, J.R., Cowen, D.: Sensing, remote Sensing of urban/suburban infrastructure and social economic attributes. Photogramm. Eng. Remote. Sens. **65**, 153–163 (2011)
2. Tran, H., Tran, T., Kervyn, M.: Dynamics of land cover/land use changes in the Mekong Delta, 1973–2011: a remote sensing analysis of the Tran Van Thoi District, Ca Mau Province, Vietnam. Remote Sens. **7**, 2899–2925 (2015)
3. Kantakumar, L., Singh, P.: Multi-temporal land use classification using hybrid approach. Egypt. J. Remote Sens. Space Sci. **18**, 289–295 (2015)
4. Sun, C., Liu, Y., Zhao, S., Zhou, M., Yang, Y., Li, F.: Classification mapping and species identification of salt marshes based on a short-time interval NDVI time-series from HJ-1 optical imagery. Int. J. Appl. Earth Obs. Geoinf. **45**, 27–41 (2016)
5. Gómez, C., White, J.C., Wulder, M.A.: Optical remotely sensed time series data for land cover classification: a review. ISPRS J. Photogramm. Remote Sens. **116**, 55–72 (2016)
6. Valero, S., Morin, D., Inglada, J., Sepulcre, G., Arias, M., Hagolle, O., Dedieu, G., Bontemps, S., et al.: Processing Sentinel-2 image time series for developing a real-time cropland mask. In: International Geoscience and Remote Sensing Symposium (IGARSS), pp. 2731–2734 (2015)
7. Nguyen, H., Doan, T., Tomppo, E., McRoberts, R.: Land use/land cover mapping using multitemporal Sentinel-2 imagery and four classification methods—a case study from Dak Nong, Vietnam. Remote Sens. **12**, 1367 (2020)
8. Maus, V., Câmara, G., Appel, M., Pebesma, E.: dtwSat : time-weighted dynamic time warping for satellite image time series analysis in R. J. Stat. Softw. **88**, 1–31 (2019)
9. Gorelick, N., Hancher, M., Dixon, M., Ilyushchenko, S., Thaus, D., Moore, R.: Google earth engine: planetary-scale geospatial analysis for everyone. Remote Sens. Environ. **202**, 18–27 (2017)

10. Rembold, F., Meroni, M., Urbano, F., Royer, A., Atzberger, C., Lemoine, G., Eerens, H., Haesen, D.: Remote sensing time series analysis for crop monitoring with the SPIRITS software: new functionalities and use examples. Front. Environ. Sci. **3**, 46 (2015)
11. Tassi, A., Vizzari, M.: Object-oriented LULC classification in Google earth engine combining SNIC, GLCM, and machine learning algorithms. Remote Sens. **12**, 3776 (2020)
12. Maxwell, A.E., Warner, T.A., Fang, F.: Implementation of machine-learning classification in remote sensing: an applied review. Int. J. Remote Sens. **39**, 2784–2817 (2018)
13. Blaschke, T.: Object based image analysis for remote sensing. ISPRS J. Photogramm. Remote Sens. **65**, 2–16 (2010)
14. Kandasamy, S., Frederic, B., Verger, A., Neveux, P., Weiss, M.: A comparison of methods for smoothing and gap filling time series of remote sensing observations: application to MODIS LAI products. Biogeosciences. **10**, 4055–4071 (2013)
15. Aryaguna, P., Danoedoro, P.: Comparison effectiveness of pixel based classification and object based classification using high resolution image in floristic composition mapping (study case: Gunung Tidar Magelang City). IOP Conf. Series Earth Environ. Sci. **47**, 012042 (2016)
16. Aggarwal, N., Srivastava, M., Dutta, M.: Comparative analysis of pixel-based and object-based classification of high resolution remote sensing images – a review. Int. J. Eng. Trends Technol. **38**, 5–11 (2016)
17. Devaux, N., Crestey, T., Leroux, C., Tisseyre, B.: Potential of Sentinel-2 satellite images to monitor vine fields grown at a territorial scale. OENO One. **53**, 51–58 (2019)
18. Ayele, G., Tebeje, A.K., Demissie, S., Belete, M., Jemberie, M., Teshome, W., Mengistu, D., Tashale, E.: Time series land cover mapping and change detection analysis Using geographic information system and remote sensing, Northern Ethiopia. Air Soil Water Res. **11**, 1178622117751603 (2018)
19. Phiri, D., Simwanda, M., Salekin, S., Nyirenda, V., Murayama, Y., Ranagalage, M.: Sentinel-2 data for land cover/use mapping: a review. Remote Sens. **12**, 2291 (2020)
20. Bishop, C.M.: Pattern Recognition and Machine Learning. Springer, New York (2006)
21. Shelestov, A., Lavreniuk, M., Kussul, N., Novikov, A., Skakun, S.: Exploring Google earth engine platform for big data processing: classification of multi-temporal satellite imagery for crop mapping. Front. Earth Sci. **5**, 17 (2017)
22. Bittencourt, H.R., Clarke, R.: Use of classification and regression trees (CART) to classify remotely-sensed digital images. IEEE Int. **6**, 3751–3753 (2003)
23. Loh, W.: Classification and regression trees. WIREs Data Min. Knowl. Discovery. **1**, 14–23 (2011)
24. Breiman, L.: Random forests. Mach. Learn. **45**, 5–32 (2001)
25. Belgiu, M., Drăguţ, L.: Random forest in remote sensing: a review of applications and future directions. ISPRS J. Photogramm. Remote Sens. **114**, 24–31 (2016)
26. Goel, E., Abhilasha, E.: Random Forest: a review. Int. J. Adv. Res. Comp. Sci. Soft. Eng. **7**, 251–257 (2017)
27. Congalton, R.: Accuracy assessment and validation of remotely sensed and other spatial information. Int. J. Wildland Fire. **10**, 321–328 (2001)
28. Congalton, R.G., Green, K.: Assessing the Accuracy of Remotely Sensed Data: Principles and Practices. CRC Press (2019)
29. Berhane, T., Lane, C., Wu, Q., Anenkhonov, O., Chepinoga, V., Autrey, B., Liu, H.: Comparing pixel- and object-based approaches in effectively classifying wetland-dominated landscapes. Remote Sens. **10**, 46 (2018)
30. Liu, D., Xia, F.: Assessing object-based classification: advantages and limitations. Remote Sens. Lett. **1**, 187–194 (2010)
31. Mutanga, O., Kumar, L.: Google Earth engine applications. Remote Sens. **11**(5), 591 (2019)

Sugarcane and Cassava Classification Using Machine Learning Approach Based on Multi-temporal Remote Sensing Data Analysis

Jirawat Daraneesrisuk [ID], Sarawut Ninsawat [ID], Chudech Losiri, and Asamaporn Sitthi

Abstract Crop identification and mapping provide valuable information about crop acreage and aid in monitoring and decision-making for government and agro-industrial businesses. Multi-temporal remote sensing is widely used for the phenological study of crops, especially in sugarcane and cassava. The development of crop growth monitoring is one of the capabilities of multi-temporal data. This study integrated the phenological characteristics from remote sensing data to develop the optimal classifier for the reliable classification of sugarcane and cassava. Sentinel-1 and Sentinel-2 images from October 2017 to September 2019 were used to classify the crop types. The random forest (RF) classifier provided the highest overall model performance accuracy than other algorithms when using the specific NDVI dataset. Furthermore, the ground truth data collected from an unmanned aerial vehicle (UAV) field survey was used to assess the classification performance and resulted in 68% accuracy for sugarcane and cassava classification. Multi-temporal remote sensing can aid in the mapping of sugarcane and cassava. The developed approach can be used for crop mapping, management, and estimation of crop production on a regional scale. Expectedly, our development could be practically adopted on a multi-crop plantation scale.

Keywords Sugarcane · Cassava · Multi-temporal · Classification · Crop phenology

J. Daraneesrisuk (✉) · S. Ninsawat
Remote Sensing and GIS, School of Engineering and Technology, Asian Institute of Technology, Pathum Thani, Thailand
e-mail: sarawutn@ait.asia

C. Losiri · A. Sitthi
Department of Geography, Faculty of Social Sciences, Srinakharinwirot University, Bangkok, Thailand
e-mail: chudech@g.swu.ac.th; asamaporn@g.swu.ac.th

© The Author(s), under exclusive license to Springer Nature Switzerland AG 2023
W. Boonpook et al. (eds.), *Applied Geography and Geoinformatics for Sustainable Development*, Springer Geography,
https://doi.org/10.1007/978-3-031-16217-6_14

1 Introduction

Agriculture is one of the essential activities that make foods for humans worldwide. Identifying crop plantation area and production statistics is vital for farming counties. Also, governments and businesses section can adapt this information to develop and plan agricultural policies for their counties, such as crop yield information. Sugarcane and cassava are the most plantation area and one of Thailand's cash crops. Also, several products of sugarcane and cassava are essential for Thailand's economy because they are the top ten ranked industrial crops. Moreover, sugarcane and cassava farming are introduced in agricultural economic zones (AEZs) by Royal Thai Government [1].

Remote sensing from space is a gainful approach that can be used for crop mapping. There are advantages to crop identification. Earth observation can be used for timely, dynamic, and large-scale studies of the land surface [2]. This approach can help to reduce time and labor. Also, it increases the accuracy, reliability, and frequency of crop information, which is valuable for designing and implementing agricultural policies [3]. Moreover, the phenology curve and pattern created by optical time-series imagery can characterize sugarcane and cassava and lead to crop type classification [4]. Also, the SAR imagery can identify the biophysical structure of sugarcane and cassava because of the unique crop canopy. Many studies used the pattern of backscatter values to classify crop types in several areas. SAR imagery can provide crop growth continuously because it can identify land surfaces under the worst weather conditions [5]. Sugarcane, cassava, and other crops have different vegetation indices and backscatter patterns; hence, crop types can be separated by different wares derived from optical and SAR images (Fig. 1).

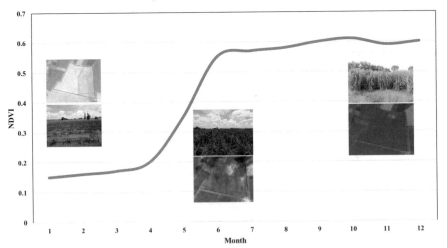

Fig. 1 The phonological curve of sugarcane plantation in the case of vegetation indices

Crop growth has periodic events in the life cycle of living species. The phase of crop plantation is different physiology. For example, sugarcane and cassava are long-time plantations that constitute a cycle of 10–12 months [6]. Also, each stage of the crop is various because these crops grow and sprout continuously. For instance, sugarcane has four main stages that lead to different physiologies: sprouting, tillering, stalk growth, and maturation [7]. Hence, the various stages of crops are not similar, and several stages of crops can show different values from remote sensing data. The selection of satellite imagery and frequency of images for crop classification depends on factors like image availability, level in crop types, and extensively of the study area [8].

This research studies classifying sugarcane and cassava areas using multi-temporal imagery of vegetation indices and backscatter coefficient. Also, remote sensing data related to the phenological patterns are analyzed by the machine learning approach to classify sugarcane and cassava. This study is separated into three parts: an analysis of vegetation indices and backscatter coefficients for sugarcane and cassava phenology; the classification of sugarcane and cassava by the optimal approach of support vector machine, random forest, and neural network algorithms; and the identification of sugarcane and cassava plantation. First, an optical and SAR imagery analysis for crop phenology, VIs, and backscatter coefficients can be generated to identify sugarcane and cassava phenological patterns. In the second part, manual labor creates training datasets based on field information from several resources such as Mitr Phol Sugarcane Company, Land Development Department (LDD), and Department of Agriculture Extension. Finally, optimal classification methods are applied in a classification approach to classify sugarcane and cassava in the study area.

2   Materials and Methods

2.1   Data Collection

The study area of this research is in Khon Kaen Province, northeast region, Thailand. The weather condition and climate properties are suitable for sugarcane and cassava plantations. Moreover, this province has approximately 982.5 and 503.8 km^2 of cultivated sugarcane and cassava. Our satellite imagery datasets were taken from European Space Agency (ESA). We have also taken the Synthetic Aperture Radar imagery of Sentinel-1 and the optical imagery of Sentinel-2 in study area scenes from 2017 to 2019. Also, samples of sugarcane and cassava farm for training, testing, and validation were collected by field survey during the crop growing season (2017–2018) from Mitr Phol Sugar Company and the Department of Agriculture Extension. Furthermore, the land use from Land Development Department (LDD) is used to identify other classes of land use such as urban areas, forests, and water bodies in the study area and to mask non-target areas after classification. In addition, unmanned aerial vehicle (UAV) images from field surveys for a simple study

Table 1 The data used for the analysis of sugarcane and cassava phonological characteristics and classification of sugarcane and cassava

No	Name	Description	Period	Source
1	Sentinel imagery	The multispectral images of Sentinal-2 level 1C and the SAR images of Sentinal-1	2017–2019	European Space Agency
2	Sugarcane farm	The shapefile of sugarcane farm boundary with location and start date of plantation	2017–2018	Mitr Phol sugar company
3	Cassava farm	The shapefile of cassava farm boundary with location and start date of plantation	2017–2018	Department of Agriculture Extension
4	Land use	The shapefile of land uses such as urban, water body, forest, etc.	2019	Land development department
5	UAV imagery	The RGB images of UAV flights from field surveys and ground trust data	Feb 2020	Field survey

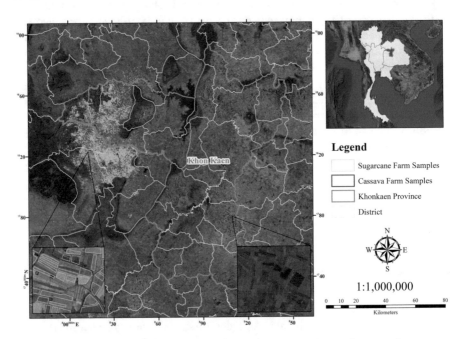

Fig. 2 The study area of Khon Kean province with sugarcane and cassava farm samples

area are used to accurately identify ground references and classified images, as shown in Table 1 (Fig. 2).

2.2 Methodology

The methodology for this study is represented in Fig. 3. The detail of this study is the crop classification using machine learning methods that integrate multi-temporal analysis.

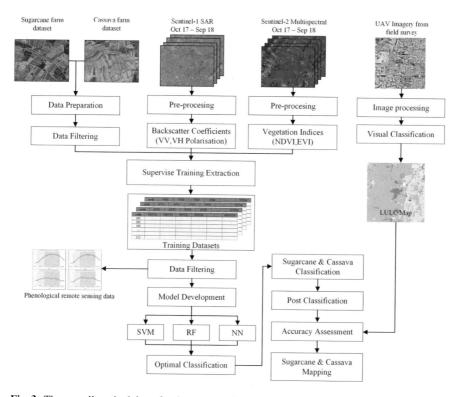

Fig. 3 The overall methodology for data preparation, model development, and classification

First, the multi-temporal imagery of Sentinel-1 SAR and Sentinel-2 optical was prepared and collected. Vegetation indices (NDVI, EVI) and backscatter values (VV, VH) were generated by raw images on the whole image between October 2017 and September 2019. In optical imagery, the raw images of Sentinel-2 were processed by geomatic and radiometric collection. Similarly, these images were generated by the NDVI and the EVI formula. Next, we used a 15-day composite approach to create periods of vegetation indices of crops. In synthetic aperture radar imagery, these images of Sentinel-1 were prepared by several methods to absolute backscatter brightness. Also, we used 15-day composite methods to generate periods of time-series data. Next, sugarcane and cassava information, including crop types, plantation, farm location, and area, were generated into vector features. Then, these features were modified by buffer techniques to reduce mix-pixel error. Finally, all vector features were extracted from each farm and averaged all pixel values in each farm. Therefore, this step generated training data for sugarcane and cassava for the classification method.

Second, the analysis of multi-temporal data for crop phenology used farm areas generated from vector features for training data. Values of vegetation indices and backscatter brightness were used by phenological analysis with statistical techniques because it can provide the reliability of results after being classified. Also, training data was required for machine learning classification methods such as support vector machine, random forest, and neural network. Then, we accurate a result of all procedures in each image dataset to determine the optimal classification approach.

The best classification method and optimal image dataset were used to classify the study area and identify sugarcane and cassava plantations. Then, the UAV images were generated into orthoimages and visually classified as the multiclass of land use and land cover. We compared the accuracy of classification performance for sugarcane and cassava. Finally, the optimal machine learning approach for sugarcane and cassava classification was used to classify the large area images. Then, we tested the accuracy assessment for sugarcane and cassava classification results.

3   Results

3.1   Sugarcane and Cassava Phenological Characteristics

We identified the phenological pattern of sugarcane and cassava through analysis of vegetation indices and backscatter coefficients from multi-temporal imagery of Sentinel-1 and Sentinel-2 satellites. The sugarcane and cassava fields are given the phenological graphs and analyzed by time-series data in 24 periods or 1 year from starting to harvesting time. The NDVI and EVI from optical images also VV and VH from SAR images were used in this analysis. Figure 4 shows the crop phenology of sugarcane (plant) in each stage. Figure 5 shows the crop phenology of sugarcane (ratoon) in each stage. Figure 6 shows the crop phenology of cassava in each stage.

3.2   Optimal Approach for Sugarcane and Cassava Classification

In this section, to start with, the results of the classification model in machine learning for sugarcane and cassava identification are briefly reported. We used three classifier models, which are support vector machine (SVM), random forest (RF), and neural network (NN), to identify the optimal classifier which appropriates for sugarcane and cassava plantation. Also, we separated four main experiments of classification that are using specific NDVI as abbreviated Exp1, using specific EVI as abbreviated Exp2, using the combination of vegetation indices (NDVI and EVI) as abbreviated Exp3, and using the combination of vegetation indices with backscatter coefficient (NDVI, EVI, VV, and VH polarization) as abbreviated Exp4. In addition, the whole dataset was filtered by linear-gap filtering method. Then, we compared the model accuracy assessment results in each experiment to show the best-performing method, suitable and effective for this study in sugarcane and cassava classification via remote sensing data.

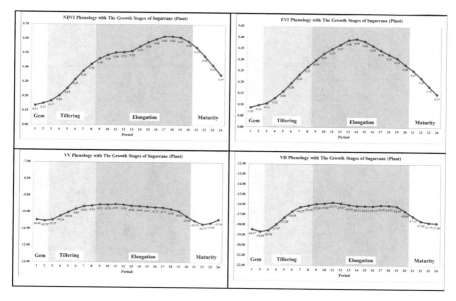

Fig. 4 The growth stage of sugarcane (plant) with the average of vegetation indices (NDVI, EVI) and backscatter values (VV, VH polarization) in 12 months (24 periods)

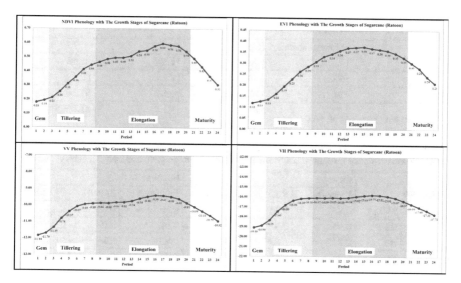

Fig. 5 The growth stage of sugarcane (ratoon) with the average of vegetation indices (NDVI, EVI) and backscatter values (VV, VH polarization) in 12 months (24 periods)

Tables 2 and 3 summarize the accuracies and performance measures of the sugarcane and cassava classification algorithms that used NDVI, EVI, a combination of vegetation indices, and a combination of vegetation indices and backscatter coefficient training datasets. The SVM provides the highest accuracy when using a

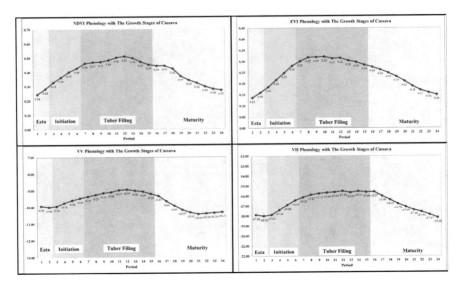

Fig. 6 The growth stage of cassava with the average of vegetation indices (NDVI, EVI) and back-scatter values (VV, VH polarization) in 12 months (24 periods)

Table 2 Accuracies of the algorithms for sugarcane and cassava classification based on NDVI, EVI, the combination of VI, and the combination of VI and SAR datasets

Classifier	Overall accuracy (%)			
	Exp1	Exp2	Exp3	Exp4
SVM	78.47	80.84	88.44	80.02
RF	96.03	87.66	93.96	91.14
NN	85.03	81.50	93.44	81.95

Table 3 Performance measurement of the algorithms for sugarcane and cassava classification

Classifier		Cassava			Sugarcane		
		Precision	Recall	F1 score	Precision	Recall	F1 score
Exp1	SVM	0.86	0.61	0.72	0.75	0.92	0.83
	RF	0.95	0.96	0.96	0.97	0.96	0.96
	NN	0.84	0.82	0.83	0.86	0.85	0.85
Exp2	SVM	0.98	0.58	0.73	0.75	0.99	0.85
	RF	0.84	0.87	0.87	0.91	0.88	0.89
	NN	0.82	0.74	0.78	0.81	0.87	0.84
Exp3	SVM	0.98	0.75	0.85	0.83	0.99	0.90
	RF	0.90	0.97	0.93	0.97	0.92	0.94
	NN	0.94	0.91	0.92	0.93	0.96	0.94
Exp4	SVM	0.71	0.93	0.81	0.93	0.69	0.79
	RF	0.85	0.98	0.91	0.98	0.86	0.92
	NN	0.77	0.85	0.81	0.87	0.79	0.83

combination of vegetation index dataset at 88.4%. The RF performs better overall accuracy when using a specific NDVI dataset at 96.03%. Finally, the NN provides the highest accuracy when using a combination of vegetation index dataset at 93.44%. Also, RF provides higher precision, recall, and F1 score than 0.95 in all classes when using a specific NDVI dataset. Therefore, we selected the random forest classifier for classification and used a specific NDVI dataset in the resulting step.

3.3 The Result of Classification

This section used the random forest classifier to classify sugarcane and cassava since it shows the best performance among the three classifiers. Moreover, the multi-temporal imagery of NDVI, which is the best dataset for evaluating between October 2018 and August 2019, was used in the survey area. Also, we prepared training datasets for this section.

The sugarcane and cassava areas were classified using multi-temporal imagery of NDVI, which is optimal classification after the evaluated performance. Figure 7 shows an obtained classification result used in October 2017–February 2018 for training data and October 2019–February 2020 for image classification. The sugarcane, cassava, and non-sugarcane/cassava of ground reference data were used to assess classification accuracy. The classification performance between a whole area of reference data from a field survey and the obtained classification result has an overall accuracy of 0.68. This performance evaluation selected all pixels in this area. In addition, the classification of the cassava area (producer's accuracy = 0.76)

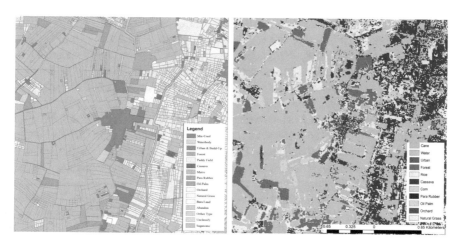

Fig. 7 The obtained result of classification by using the optimal classification approach

Table 4 The accuracy assessment of classification performed from multi-temporal of NDVI imagery

		Reference data (field survey)			Total (pixel)	User's accuracy
		Non-sugarcane/cassava	Cassava	Sugarcane		
Classified data	Non-sugarcane/cassava	**9010**	94	11,643	20,747	0.43
	Cassava	3290	**1159**	2154	6603	0.18
	Sugarcane	4988	265	**37,089**	42,342	0.88
	Total (pixel)	17,288	1518	50,886	**69,692**	
	Producer's accuracy	0.52	0.76	0.73		**OA = 0.68**

was better compared to the sugarcane area (producer's accuracy = 0.73) and non-sugarcane and cassava area (producer's accuracy = 0.52), shown as Table 4.

4 Discussion

The vegetation indices and the backscatter coefficients can be used to identify and analyze the characteristic of sugarcane and cassava phenology in multi-temporal remote sensing data. To benefit from the multi-temporal information, the data of crop growing is one of the most capabilities of remote sensing data throughout the plantation period. This study provides the difference in phenology between sugarcane and cassava in multiple cases of remote sensing data such as optical and SAR information. Also, applying the pseudo-15-day composite approach can reduce the missing values from problems of satellite imagery to increase a better quality of vegetation index and backscatter coefficient data for the analysis of phenology from Sentinel-1 and Sentinel-2 products released from 2017 to 2018. However, this research has insufficient plantation data on cassava, which can decrease the accuracy of the obtained results. Moreover, this research divides into plant sugarcane, ratoon sugarcane, and cassava. Thus, there is a general phenological curve, not separate subspecies of these crop types.

In this research, the supervised training dataset was selected for 5 months because this period can provide an understanding of the difference of crop types between sugarcane and cassava plantations. The random forest classifier is the optimal method for sugarcane and cassava classification, providing higher accuracy than the support vector machine and the neural network classifier. This approach was the best algorithm when they used the NDVI datasets better than other datasets in multi-temporal data. Alternatively, the case of the SVM and neural network classifiers provides the highest accuracy of model performance when it uses the combination of vegetation index (NDVI+EVI) dataset. Also, the whole training dataset was filtered by many conditions for sugarcane and cassava to increase the performance of the classification results. As a result, various classifiers can provide a few differences in overall accuracy between 78% and 96% when we used to classify a sugarcane and cassava phenology.

5 Conclusion and Recommendation

In brief, the classification of sugarcane and cassava by using phenological patterns based on the machine learning approach is used in this study. The supervised training dataset was used to evaluate the performance of classifier algorithms (SVM, RF, and NN). Initially, four training datasets for classification were improved in quality by gap-filtering methods. Also, the reference dataset was split into training (70%) and testing (30%). The random forest classifier is the optimal approach for this study, providing higher accuracy of classification performance models than other classifiers. The findings from this research have demonstrated the successful application of time-series data for collective mapping of sugarcane and cassava fields using the information on crop phenology in this study area. The profiles of these crops were essential for developing a phenology-based approach for sugarcane and cassava mapping.

This study also unveils many topics worthy of further research efforts. The recommended future works are provided below to improve the performance of sugarcane and cassava classification-based multi-temporal remote sensing data. These recommendations can also be implemented to classify other crop types and cover a large spatial extent. Although the generation of phenological characteristics using remote sensing data showed improvement of results, it still requires accurate plantation date information to generate crop phenology. Thus, researchers who need to solve this problem can use several models to create a phenological curve. Also, this research used a single crop cycle of sugarcane and cassava; thus, considering multiple crop cycles from a different year is more accurate. Moreover, the variance of cassava phenology was occured because the cassava information in this study was insignificant data less than sugarcane. In the case of different species of sugarcane and cassava, the separation of sub-spicy types of sugarcane and cassava can improve unique phenology.

References

1. Boonyanam, N.: Agricultural economic zones in Thailand. Land Use Policy. **99**, 102774 (2020)
2. Xu, J., Meng, J., Quackenbush, L.J.: Use of remote sensing to predict the optimal harvest date of corn. Field Crop Res. **236**, 1–13 (2019)
3. Arias, M., Campo-Bescós, M.Á., Álvarez-Mozos, J.: Crop classification based on temporal signatures of sentinel-1 observations over Navarre Province, Spain. Remote Sens. **12**(2), 278 (2020)

4. Ouzemou, J.-E., et al.: Crop type mapping from pansharpened Landsat 8 NDVI data: a case of a highly fragmented and intensive agricultural system. Remote Sens. Appl. Soc. Environ. **11**, 94–103 (2018)
5. Clauss, K., et al.: Estimating rice production in the Mekong Delta, Vietnam, utilizing time series of Sentinel-1 SAR data. Int. J. Appl. Earth Obs. Geoinf. **73**, 574–585 (2018)
6. Azar, R., et al.: Assessing in-season crop classification performance using satellite data: a test case in northern Italy. Eur. J. Remote Sens. **49**(1), 361–380 (2016)
7. Santos, F., Diola, V.: Chapter 2 – Physiology. In: Santos, F., Borém, A., Caldas, C. (eds.) Sugarcane, pp. 13–33. Academic Press, San Diego (2015)
8. Saini, R., Ghosh, S.: Crop classification on single date SENTINEL-2 imagery using random forest and SUPPOR vector machine. ISPRS Int. Arch. Photogramm. Remote. Sens. Spat. Inf. Sci. **XLII-5**, 683–688 (2018)

Google Earth Engine Algorithm for Evaluating the Performance of Landsat OLI-8 and Sentinel-2 in Mangrove Monitoring

Asamaporn Sitthi

Abstract Mapping mangrove forest extents is important to assess the general health of coastal ecosystems. However, as mangrove forests are dense by nature and tend to grow in mudflats, this chapter introduces physical challenges that often hinder remote scientists from easily accessing the mangrove forestry in field surveys. The publicly available imagery produced by the commonly used Landsat OLI-8 satellite imagery and the newly released Sentinel-2 satellites may be used to map mangrove forest extents remotely. This chapter uses the Google Earth Engine (GEE) tool to conduct a comparative evaluation of the performance of Landsat OLI-8 and Sentinel-2 satellite imagery to map mangrove forest extents on the coast of the Trat province of Thailand. The results indicated that Sentinel-2 is visually and quantitatively preferable compared to Landsat OLI-8 when mapping mangrove forest extents with 88 percentage of accuracy.

Keywords Mangrove · Remote sensing · Artificial intelligence · Machine learning

1 Introduction

Mangroves are essential to the ecosystems in which they exist. They can support biodiversity by providing habitats for other organisms, their dense roots maintain water quality by trapping pollutants, they stabilize shorelines by slowing erosion, they support the fishing industry, and they sequester more tonnes of below-ground carbon per hectare than boreal, temperate, and tropical forests combined [1]. Consequently, remote scientists use satellite imagery to map mangrove forest extents and temporal changes [3]. Several researches studied mangrove

A. Sitthi (✉)
Department of Geography, Faculty of Social Science, Srinakharinwirot University, Bangkok, Thailand
e-mail: asamaporn@g.swu.ac.th

© The Author(s), under exclusive license to Springer Nature Switzerland AG 2023
W. Boonpook et al. (eds.), *Applied Geography and Geoinformatics for Sustainable Development*, Springer Geography,
https://doi.org/10.1007/978-3-031-16217-6_15

monitoring in terms of changes, rehabilitation, and regeneration in a number of applications [3]. Quantitative changes are normally collected at multi-scales. The results of which are commonly used to conduct time-series analyses of mangrove forests in a given area. The publicly available imagery released by Landsat OLI-8 is commonly used to map mangrove forest extents. However, the medium-resolution imagery may not provide sufficient detail to classify mangrove forestry with a high level of accuracy [6]. The Sentinel-2A and Sentinel-2B satellites, launched in 2015, offer the potential to provide greater accuracy when mapping mangrove extents, as the red, green, blue, and near-infrared bands have spatial resolutions of 10 meters and the three red-edge bands allow for more accurate vegetation detection [2].

2 Study Area and Dataset

The coast of the Trat province, located in eastern Thailand, is chosen as the study area to serve as preparation for additional research in this study area. An 850-meter buffer is placed around the mangrove forestry in order to map the surrounding LULC classes in the region. The images of the region are from 2017 (Fig. 1).

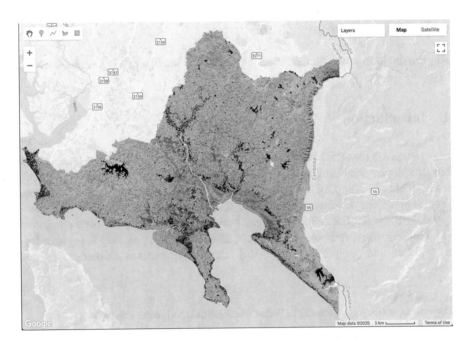

Fig. 1 Satellite imagery covers the study area

3   Methodology

Landsat OLI-8 and Sentinel-2 imagery are used within the GEE to determine spectral profiles; perform unsupervised classifications; generate stratified random coordinates; 70% of which are for training and 30% for validation; train the data using the random forest classifier; assess the accuracy of the results through post-classification confusion matrix and error matrix computations; and determine the area in square meters of which each class is composed, as well as their respective proportions of the total study area. The spectral profiles are used throughout the classification process to compare the reflectances of a given pixel or cluster of pixels to the typical values for each of the four classes considered in this report (Fig. 2).

3.1   Spectral Signature Analysis

3.1.1   Landsat OLI-8 vs Sentinel-2A

The spectral signature analysis will be used to classify that area. The indicator traits associated with mangrove reflectance must be investigated. The spectral signature is based on data from optical satellite imagery to identify and compare the reflectance of different objects. Depending on the composition, they have similar reflectance values, which influence land cover reflectance. The USGS Landsat 8 Collection 1 Tier 1 TOA Reflectance is used within the Google Earth Engine to represent a

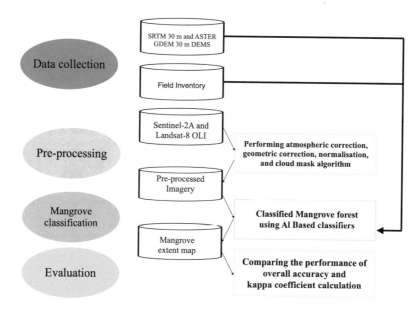

Fig. 2 Methodological flow of the research work

common source of satellite data that remote scientists use to map mangrove forestry. This source in particular is used because Landsat OLI-8's raw digital numbers (DN) have already been converted to TOA reflectances, and the source offers FMask functionality to remove cloud cover. Landsat OLI-8 is a convenient source of satellite imagery, as it is free to the public and information and documentation regarding the satellite are widely accessible. The satellite has been releasing imagery since 2013 and thus can be used as a resource for temporal analyses of mangrove forestry. Nine of Landsat OLI-8's bands have resolutions of 30 meters, and two bands – Band 8 (Panchromatic) and Band 9 (Cirrus) – have resolutions of 15 meters [2]. This limits the detail that remote scientists can use to identify mangrove forestry in supervised classifications using Landsat OLI-8 imagery. Cloud masking is performed on Landsat OLI-8 data using the "simpleCloudScore" function in GEE. The spectral profiles associated with the wavelengths of the bands of Landsat OLI-8 images are estimated using the median reflectances from sample pixels in each class in the study area (Fig. 3).

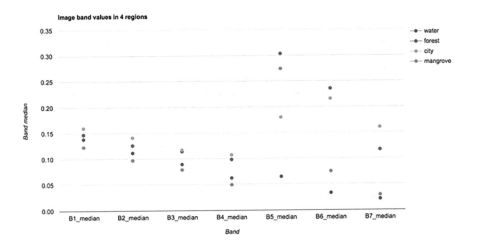

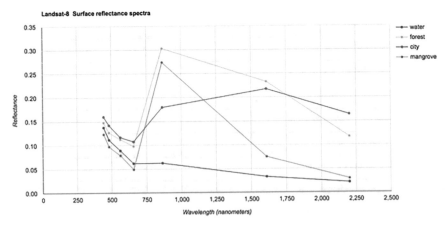

Fig. 3 Spectral profiles of each class in Landsat OLI-8 imagery

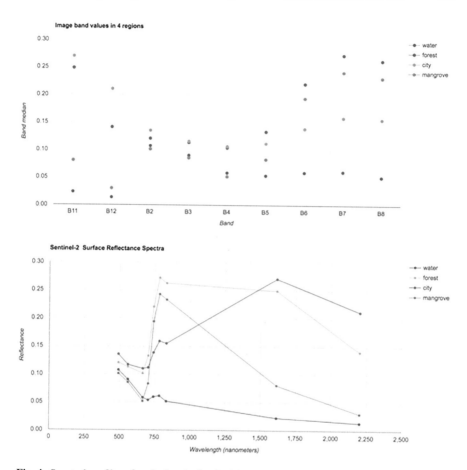

Fig. 4 Spectral profiles of each class in Sentinel-2 imagery

As a recently released satellite, the documentation available for Sentinel-2 imagery is temporarily limited, and the satellite cannot offer data prior to its release in 2013 (Fig. 4).

3.2 Training and Validation Data

Unsupervised classifications are used on both Landsat OLI-8 satellite imagery and Sentinel-2 imagery to determine appropriate LULC classes based on the reflectances shown in the study area. Four main clusters of reflectance values were found; as a result, agriculture, urban/bare land, mangrove forestry, and water bodies/shrimp farms are the four classes considered in the study. Urban areas and bare land are placed in the same class, as urban and bare land each make up small percentages of the total LULC of the study area, and they share similar reflectance values. Similarly, water bodies and shrimp farms have similar reflectance values and share a small proportion of the LULC of the Trat coast (Fig. 5).

Stratified random sampling is used to generate 414 random points within the study area [5]. The number of points in each class is proportional to the amount [2]. However, the satellite data may be more fit for mangrove extent mapping, as its bands have spatial resolutions of 10 meters for Bands 2–4, Band 8, and Band QA10; 20 meters for Bands 5–7, Band 8A, Bands 11–12, and Band QA20; and 60 meters for Band 1, Bands 9–10, and Band QA60 [2]. These allow for Sentinel-2 to release high-resolution images. Sentinel-2 MSI, MultiSpectral Instrument, Level-1C is used as the source of Sentinel-2 imagery in GEE. It represents the TOA reflectance, scaled by 10,000. For this reason, reflectances are divided by 10,000 before they are assessed. Cloud masking is performed on Sentinel- 2 data using the cloud and cirrus bits of the QA band in Sentinel-2. The median pixels are then chosen from the resulting cloud-free image, as shown in Fig. 1. Similarly, to Landsat OLI-8, the spectral profiles associated with the wavelengths of the bands of Sentinel-2 images are estimated using the median reflectances from sample pixels in each class in the study area. One hundred and fifty random points are allocated for agriculture, 50 points are given to urban/bare land, 164 points represent mangroves, and 50 points are provided for water bodies and shrimp farms. Seventy percent of the points in each class are used to serve as training data for the classification process, and the remaining 30% are saved to be used for validation. The stratified random points are used as parameters in a JavaScript function that randomly selects 70% of the pixels to be used as training data and 30% to be used as validation data. The remaining 30% of the stratified random points are used to validate the data with the same

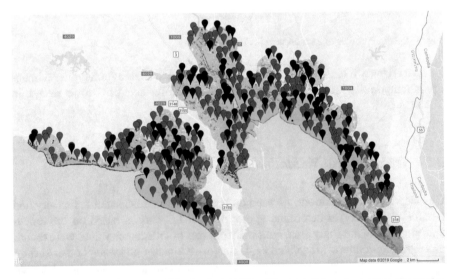

Fig. 5 Imagery of the Trat coast of Thailand with 70% of the total 414 random points allocated to training data (red) and 30% to be used for validation (black) in the GEE

classifier used to develop the classifications. The results of this will determine the accuracy of the RF classifier, which can be used to produce a confusion matrix that indicates the overall accuracy of the training using the RF classifier, as well as the validation error.

3.3 AI-Based Classification

Training samples for each class are collected as polygons in GEE engine. The random forest (RF) supervised classification method is used to classify the regions within the study area. Random forest classification uses bootstrap aggregation to classify the data using training data. When tested against other commonly used algorithms for land cover classification, random forest classification outperforms all of the other classification methods, including maximum likelihood, minimum distance, decision trees, neural networks, and support vector machine (SVM) classifiers [4]. For this reason, and due to its popularity in remote sensing applications, random forest classification is used within GEE to classify the study area. The RF classifier is trained using 10 trees with bands and the randomly chosen 70% of the stratified random points.

3.4 Post-classification Accuracy Assessment Result and Discussion

Confusion matrices are constructed to assess the accuracy of the training and the validation error. Tables 1 and 2 highlight the confusion matrix for validation error for the Landsat OLI-8 and Sentinel-2 mosaics, respectively. The number of pixels

Table 1 Landsat OLI-8 re-substitution error matrix

Classification	Landsat OLI-8 re-substitution error matrix			
Agriculture	10,066	0	14	0
Bare land/urban	4	516	0	1
Mangrove	6	0	15,464	0
Water body/shrimp farm	6	0	0	1040

Training overall accuracy: 0.998893642

Table 2 Sentinel-2 re-substitution error matrix

Classification	Sentinel-2 re-substitution error matrix			
Agriculture	12,408	3	2	1
Bare land/urban	13	4037	0	0
Mangrove	2	0	69,654	2
Water body/shrimp farm	1	0	9	4449

Training overall accuracy: 0.999635685

that belong to each class in the resulting classified layer is used to calculate the area of each class in the study area. The areas are then used to determine the proportion of area in the coastal Trat province that belongs to agriculture, urban/bare land, mangrove forestry, and water bodies/shrimp farms. In order to make the numbers of pixels in each class in the Landsat OLI-8 and Sentinel-2 images comparable, Landsat pixels were multiplied by 900 meters^2 to find the square meters of land use and land cover per class. This same process was completed for Sentinel-2, but with 100 meters^2.

4 Result

Sentinel-2 and Landsat OLI-8 satellites are both able to classify mangrove forestry with acceptable levels of error; however, the high spatial resolutions and red-edge bands allow for Sentinel-2 to be particularly accurate for vegetation analysis, and this includes mangrove extent mapping.

4.1 Visual Interpretation

The cloud-free mosaic with Landsat OLI-8 (left) and Sentinel-2 (right) images is shown in Fig. 6. Though both images indicate differences in vegetation reflectances between mangrove forestry, healthy agriculture, and bare agricultural land, Landsat OLI-8 does not show as significant of a visual difference between healthy vegetation and unhealthy or off-season agricultural regions. As such, it is simpler to classify small clusters of pixels using the Sentinel-2 image, from a visual standpoint. Both mosaics distinctly show water bodies and urban or bare land areas.

4.2 Accuracy Assessment

The accuracy of classification training through re-substitution error matrices is shared in Tables 1 and 2. The user's and producer's accuracy are comparable between Landsat OLI-8 and Sentinel-2 satellite imagery.

As for the validation error matrices shown in Tables 3 and 4, Sentinel-2 performs slightly more accurately than the Landsat OLI-8 image, with overall validation error accuracies of 0.88 and 0.872; this is likely to be due to the higher spatial resolution of Sentinel-2 imagery. Spatially small features such as roads or streams can be detected by Sentinel-2 data, yet they are likely to be too small in size to take up the majority of a 900-square meter area; thus, their reflectances will not be represented in a given pixel in Landsat OLI-8 satellite imagery. The slight difference in validation accuracy is likely to be due to more accurate classifications, which are thanks to the greater amount of detail shown in a given area in Sentinal-2 imagery.

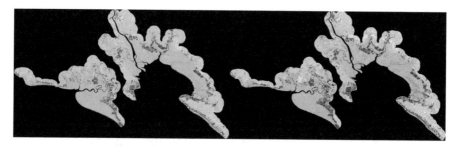

Fig. 6 Imagery of the Trat coast of Thailand from Landsat OLI-8 (left) and Sentinel-2 (right) in vegetation analysis (RGB: SWIR 1, NIR, red) composite

Table 3 Landsat OLI-8 validation error matrix

Classification	Landsat OLI-8 validation error matrix			
Agriculture	45	0	0	0
Bare land/urban	9	5	0	1
Mangrove	6	0	44	0
Water body/shrimp farm	0	0	0	15

Validation overall accuracy: 0.87

Table 4 Sentinel-2 validation error matrix

Classification	Sentinel-2 validation error matrix			
Agriculture	43	1	1	0
Bare land/urban	6	5	0	4
Mangrove	3	0	47	0
Water body/shrimp farm	0	0	0	15

Validation overall accuracy: 0.88

Figure 7 shows the LULC-classified Trat coast based on Landsat OLI-8 (a) and Sentinel-2 (b) imagery. It is immediately noticeable, both from the figures and the validation error matrices, that Landsat OLI-8 satellite data has classified very little of the study area as urban or bare land. Many of these values that are bare land were expected to be agriculture, and vice versa, as shown in the error matrices; this is largely due to the supervisor's interpretation of bare land and ways in which bare land varies from inactive, unused, or weak agricultural land in supervised classifications. The discrepancy may be more severe for Landsat OLI-8 as it has a lower spatial resolution, which causes small variations in vegetation health to be less recognizable in Landsat OLI-8 imagery. This also may be a factor behind the difference in LULC area between bare land and agriculture in Table 5.

Because Sentinel-2 produces high-resolution images, small features in the land use and land cover are recognizable in the image; for instance, more roads can be seen in the LULC-classified image using Sentinel-2 imagery; the thin red lines indicate that the roads impact the spectral reflectance individual Sentinel-2 pixels, and the random forest classification recognizes the roads that appeared in the training

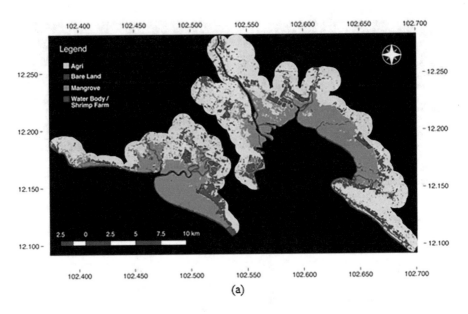

Land Use and Land Cover in 2017

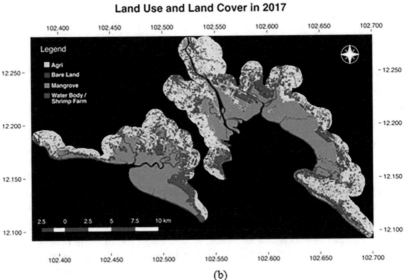

Fig. 7 Random forest-classified imagery of the Trat coast of Thailand from Landsat OLI-8 (**a**) and Sentinel-2 (**b**) in 2017

data. Similarly, the Sentinel-2 classified image includes clear details of the features on the land; thin patches of agricultural land are detected in between shrimp farms, and small streams are detected among the mangrove forestry. There may be more urban and bare land represented in the Sentinel-2 image due to the fact that the training data included a greater variety of examples of urban and bare land, as they were more recognizable and identifiable as urban or bare land use and land cover.

Table 5 Landsat OLI-8 (top) and Sentinel-2 (bottom) LULC area and proportion of total land cover in the region of interest

Classification (Landsat8)	Pixels	Area (m²)	Area percentage
Agriculture	137,265	123,538,471	53.01
Bare land/urban	7179	6,461,548	2.77
Mangrove	80,834	71,751,457	31.22
Water body/shrimp farm	33,643	30,279,017	12.99
Classification (Sentinel-2)	**Pixels**	**Area (m²)**	**Area percentage**
Agriculture	922,930	92,293,051	39.60
Bare land/urban	200,956	20,095,614	8.62
Mangrove	837,939	83,793,987	35.95
Water body/shrimp farm	368,896	36,889,680	15.83

5   Concluding Remarks

The extra red-edge bands in the Sentinel-2 imagery permit for the images to be more sensitive to variations in vegetation, which allows for mangrove forestry to be distinguished from other types of forest, agriculture, and other foliage. Although both Sentinel-2 and Landsat OLI-8 were able to determine mangrove extents and distinguish mangrove forestry from other forms of vegetation, Sentinel-2's high-resolution imagery detected and included smaller features on the land, which resulted in a sharper and more defined classification map of the mangrove forest extent and the surrounding land use and land cover areas in Trat.

References

1. Alongi, D.M.: Carbon sequestration in mangrove forests. Carbon Manag. **3**(3), 313–322 (2014). https://doi.org/10.4155/cmt.12.20
2. Gorelick, N., Hancher, M., Dixon, M., Ilyushchenko, S., Thau, D., Moore, R.: Google earth engine: planetary-scale geospatial analysis for everyone. Remote Sens. Environ. **202**, 18–27 (2017)
3. Kuenzer, C., Bluemel, A., Gebhardt, S., Quoc, T.V., Dech, S.: Remote sensing of mangrove ecosystems: a review. Remote Sens. **3**, 878–928 (2011). https://doi.org/10.3390/rs3050878
4. Kulkarni, A.D., Lowe, B.: Random Forest Algorithm for Land Cover Classification. Computer Science Faculty Publications and Presentations. Paper 1 (2016) http://hdl.handle.net/10950/341
5. Pimple, U., Simonetti, D., Sitthi, A., Pungkul, S., Leadprathom, K., Skupek, H., Som-ard, J., Gond, V., Towprayoon, S.: Google earth engine based three decadal landsat imagery analysis for mapping of mangrove forests and its surroundings in the Trat Province of Thailand. J. Comput. Commun. **6**, 247–264 (2018). https://doi.org/10.4236/jcc.2018.61025
6. Wang, D., et al.: Evaluating the performance of Sentinel-2, Landsat 8 and Pléiades-1 in mapping mangrove extent and species. Remote Sens. **10**(9), 1468 (2018). https://doi.org/10.3390/rs10091468

Estimation of Aboveground Biomass and Carbon Stock Using Remote Sensing Data in Sakaerat Environmental Research Station, Thailand

Sinlapachat Pungpa ⓘ, Sirilak Chumkiew ⓘ, and Pantip Piyatadsananon ⓘ

Abstract Forests play various crucial roles in ecosystems, particularly reducing carbon dioxide in the atmosphere contributed by anthropogenic activities. The biomass assessment contributes to the comprehension of forest system changes and the forest's carbon absorption. According to the limitations of manual biomass measurements, remote sensing (RS) has been broadly applied to estimate biomass in recent years. This study aims (1) to examine an index obtained from satellite images highly correlated with the aboveground biomass (AGB) in Sakaerat Environmental Research Station (SERS) and (2) to estimate the AGB and carbon stock around the station, which is dominated by extensive fertile forest. Regression analysis presented the relationship between AGB and carbon stock using remote sensing data from Landsat images and their indices. The NDVI provides the highest correlation for generating the AGB model. As a result, the AGB was 7.70 tons ha^{-1}, corresponding to a total carbon stock of 3.62 tons ha^{-1}. It clearly shows that the amount of AGB is directly related to the amount of carbon stock, which is also varied by forest types. However, as the limitation of medium satellite images of Landsat, a higher resolution of satellite images, such as Sentinel-2, would suggest more accurate carbon stock estimation. Furthermore, the available sequence fieldwork data of AGB, which are distributed over the study area, would increase the accuracy of the result calculated from the remote sensing technique.

Keywords Aboveground biomass · Remote sensing · Sakaerat Environmental Research Station · Carbon stock estimation

S. Pungpa · S. Chumkiew (✉)
School of Biology, Institute of Science, Suranaree University of Technology,
Nakhon-Ratchasima, Thailand
e-mail: s.chumkiew@sut.ac.th

P. Piyatadsananon
School of Geoinformatics, Institute of Science, Suranaree University of Technology,
Nakhon-Ratchasima, Thailand

W. Boonpook et al. (eds.), *Applied Geography and Geoinformatics for
Sustainable Development*, Springer Geography,
https://doi.org/10.1007/978-3-031-16217-6_16

1   Introduction

Global warming and climate change have been the main consequence of anthropogenic activities over the past few decades. These phenomena are mainly due to the dramatic increase in greenhouse gas emissions since the nineteenth century and the highest record in 2019 [1]. The main contributor is carbon dioxide (CO_2) [2]; between 1970 and 2002, global CO_2 increased up to 70% [3]. Higher CO_2 emissions occurred with several causes, such as deforestation, inappropriate agricultural practices, and soil degradation [3, 4]. Forests reduce the atmospheric CO_2 by photosynthesis through plants; consequently, the carbon stock of forest patches is essential for environmental planning and management related to global climate.

The forest tree height and AGB are essential to understanding the global carbon cycle and the functioning of the economic mechanisms to protect and enhance forest carbon stocks [5]. AGB includes all vegetation above ground (i.e., stems, branches, bark, seeds, flowers, and foliage of living plants), with carbon accounting for 50% approximately of its composition [6]. Certainly, the most popular and accurate result is a traditional measurement based on fieldwork; however, this measurement is difficult to conduct over extensive forests and costly, time-consuming, and labor-intensive [7, 8]. Previous studies focused mainly on plant biomass and carbon stock estimation employing traditional methods [9–11]. As remote sensing (RS) has become more applicable for forest research in these recent years, various satellite indices have correlated with plant biomass to gain the mass quantity [12, 13]. Using RS to investigate the forest biomass would lower the difficulties of field measurements and time constraints.

SERS, a station covered with large patches of forests, is located in Nakhon Ratchasima province, northeast Thailand. SERS was established in 1967 and administered by the Thailand Institute of Scientific and Technological Research (TISTR) to facilitate ecological and environmental research [14]. The two forest types, dry dipterocarp forest (DDF) and dry evergreen forest (DEF), cover 70% of the station area [14]. There are bamboo, plantation forests, and grasslands in the rest of the station area. Both DDF and DEF are generally found in northeastern and other regions of Thailand. However, DDF in SERS was claimed as the most fertile DDF forest in the country.

This study aims to develop a technique for assessing AGB and carbon stock in forest patches. The proper index was investigated to estimate AGB and carbon stock in SER by (1) comparing the relationship between various indices from satellite imagery with AGB obtained from allometric equations and (2) estimating AGB and carbon stock in the research station. For the results, it would provide a potential tool for estimation of AGB and carbon stocks according to two major forest types, DDF and DEF, using RS images.

2 Methodology

A series of satellite images of Landsat 5 and Landsat 9 were downloaded from the USGS website. Geoprocessing is operated through ArcGIS 10.3 program. The satellite images, Landsat 5 and Landsat 9, were selected by the compatible duration between both satellite's designed life expectancy and fieldwork collection. These images present the SERS' land-use and land cover changes over time. Land-use and land cover classification were done by supervised classification with fieldwork validations. Several indices generated from the satellite data were also examined and validated before using them for AGB estimation. Regression analysis was used to estimate AGB and carbon stock in this study through two variables (AGB in study plots and satellite indices from Landsat 5 data). The correlations were compared from each simple regression model to obtain a proper model (determined from coefficient of determination: R^2) for estimating AGB and carbon stock. The model was used to estimate SERS's current AGB and carbon stock based on Landsat 9 satellite. Carbon stock was calculated with the biomass multiplied by the default value (0.47) [6].

2.1 Study Area

The study area is located in SERS, Nakhon Ratchasima province, northeast Thailand (see Fig. 1). The station covers up to 7796.24 hectares (between latitude 14°26′–14°33′ N and longitude 10°50′–101°57′ E). The two majority forests, DEF (5275.63 hectares) and DDF (2443.52 hectares), are generally found in the station. The shapefiles of the study area of the SERS were retrieved from the Lower Northeastern Regional center of Geo-Informatics and Space Technology Development Agency (LNE-GISTDA).

2.2 Field Data Survey

Inventory data of 2-year permanent plots in 2006 and 2009 used in this study was collected by researchers [23] and applied for the validation in this study (see Fig. 1). The data were collected from April to July 2006 and 30 January to 5 February 2009. There are 25 (the major is DEF) and 33 plots (the major is DDF) in the data in 2006 and 2009, respectively.

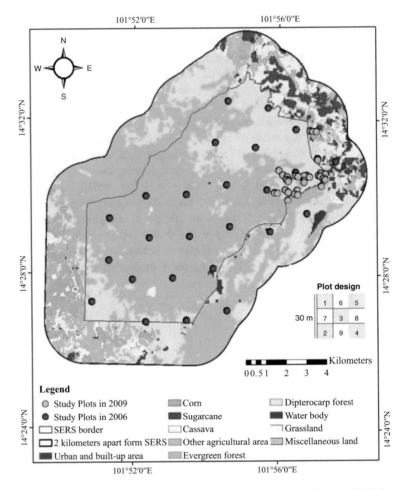

Fig. 1 Inventory map and study plot design, yellow highlights in the study area of SERS

2.3 Satellite Data

Landsat satellite images were obtained from the United States Geological Survey (USGS). This study includes images from Landsat 5 TM and Landsat 9 OLI-2 with a spatial resolution of 30 × 30 m (Path 128/Row 050). The selected satellite images were considered cloud-free images both in 2006 and 2009. The AGB and carbon stock in 2006 and 2009 were calculated from the satellite images and indices from satellite data with a consideration of the regression model. The regression model was used to estimate AGB and carbon stock in 2022 regarding the accuracy of the fieldwork.

2.4 Index Calculation

This study examines four indices – NDVI, SAVI, GNDVI, and EXG – calculated using satellite images and data. The indices used in this study were reviewed by considering the previous studies [12, 13]. The chosen index that performs the highest R^2 with AGB through the regression model was chosen as the proper index for this study.

2.5 AGB Estimation in Study Plots

AGB in study plots was estimated with allometric equations to generate regression models with satellite indices. Inventory data include only one variable, diameter at breast height (D) of trees, which can still be used to estimate biomass. The two allometric equations were selected to include only D as the variable and compared in this study.

Allometric equations used in this study:

$$\ln(AGB) = -2.286 + 2.471\ln(D) \ [16]$$

(1)

$$AGB = 38.4908 - 11.7883D + 1.1926D^2 \ [17]$$

(2)

Abbreviation:

ln: natural logarithm
D: diameter at breast height (cm)

3 Findings

3.1 Regression Model

The optimization regression model includes the variables of NDVI (dependent variable) and AGB by Eq. (2) studied in SERS. R^2 is closed to 0 in all regression models (0.01–0.05) in this study (see Table 1). NDVI is the effective index for estimation of AGB in SERS that is mainly dominated by evergreen and dipterocarp forests.

A linear regression equation is:

$$NDVI = 0.3573 + 0.00002363(AGB); r = 0.22, \ R^2 = 0.05$$

(3)

Table 1 Relationship between satellite indices and AGB in study plots

Year	Index	Equation (1)		Equation (2)	
		r	R^2	r	R^2
2006	NDVI	0.115222269	0.013276171	0.143527439	0.020600126
	SAVI	0.114954356	0.013214504	0.143242143	0.020518311
	EXG	0.136969334	0.018760598	0.173034712	0.029941011
	GNDVI	0.095784485	0.009174668	0.123647959	0.015288818
2009	NDVI	0.201885801	0.040757877	0.214978988	0.046215965
	SAVI	0.201853834	0.04074497	0.214951437	0.04620412
	EXG	0.1596605	0.025491475	0.176286481	0.031076923
	GNDVI	0.199541394	0.039816768	0.211310292	0.04465204

Table 2 Current AGB and carbon stock in SERS

Forest/area type	AGB (tons ha^{-1})	Carbon stock (tons ha^{-1})
DDF	4.52	2.12
DEF	9.25	4.34
All SERS territory	7.70	3.62

3.2   AGB and Carbon Stock in SERS

AGB and carbon stocks are 7.70 and 3.62 tons ha^{-1}, respectively, in SERS territory in the current year (see Table 2; Fig. 2). There are AGB and carbon stocks of 4.52 and 2.12 tons ha^{-1}, respectively, in DDF and 9.25 and 4.34 tons ha^{-1}, respectively, in DEF. The estimated AGB and carbon stock in the same years as previous studies (1993 and 2004) operated with the traditional method to validate model accuracy. Employing the regression model generated in this study, SERS showed an AGB and carbon stock of 76.60 and 26.00 tons ha^{-1}, respectively, in DEF in 1993 and 47.03 and 22.10 tons ha^{-1}, respectively, DDF in 2004.

4   Discussion and Conclusion

4.1   NDVI as an Optimal Index in AGB Estimation

NDVI is effective for AGB estimation in SERS by generating a simple linear regression model. The correlation coefficient between NDVI and AGB in the optimal model equals a previous study that used Quickbird image ($r = 0.21$) [17]. The results support the use of NDVI to estimate plant biomass in other studies [18–20]. For NDVI, the TM band 4 (near-infrared band) is the only factor to identify vegetation change for the forest patches with higher crown closure because of the absorption of intense chlorophyll-a in TM band 3 (red band) [21]. For this reason, it presents the NDVI as an optimal index among four different indices tested in SERS. Because the NDVI often approaches saturation level due to high crown closure, it is suitable

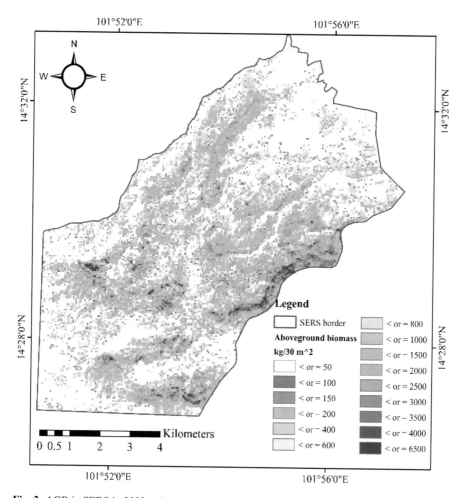

Fig. 2 AGB in SERS in 2022 estimated by applying remote sensing data

to be used in forest patches with medium crown closure [21]. It can be concluded that the use of NDVI is also based on crown closure classification.

4.2 Limitation

The limitations of this study are both satellite imagery and fieldwork methods in inventory data. The limitations of satellite imagery are (1) the different dates between satellite imagery and field data collection, (2) unavailable images due to the high percentage of cloud coverage, and (3) the low resolution of satellite imageries, which were used to generate regression models. Additionally, the limitations of the fieldwork methods are as follows: (1) The inventory data applied in this study limits the use of specific allometric equations due to the lack of a variable (tree height), (2) the plot design is not the same size and location as the image pixel, which contributes error in the relation between satellite indices and plant biomass. Further studies could conduct AGB estimation using remote sensing by a high-resolution satellite (such as Sentinel-2) consistent with the study plot design.

4.3 Traditional and RS Method on AGB Comparison

Employing the RS technique shows lower results for AGB and carbon stocks than the traditional method in SERS. In 1993, in DEF, AGB and carbon stocks were 76.60 and 26.00 tons ha^{-1} as estimated by RS, whereas they were 424.9 and 212 tons ha^{-1} approximately as estimated by manual measurement [10]. In 2004, in DDF, AGB and carbon stocks were 47.03 and 22.10 tons ha^{-1} as estimated by RS, whereas AGB and carbon stock were 89.96 and 45.58 tons ha^{-1}, approximately as estimated manually [11]. RS, used in this study, provides 1.91 times lower AGB and carbon stock in DDF and 5.39 times lower in DEF than traditional methods. DEF shows a higher AGB than DDF, corresponding to the previous study in northeastern Thailand [22]. Even the RS method presents different AGB amounts compared with a traditional method.

4.4 Practical Implication

The findings indicate that the model in this study can be used to separate AGB by forest types (DDF and DEF). RS methods need more improvement to reach manual measurement standards to gain the actual amount of AGB and carbon stock. Applying RS in plant biomass estimation would be more suitable for large area scales due to its low cost, time consumption, and less labor.

Acknowledgments This work is supported in part by the Development and Promotion of Science and Technology Talents Project (DPST), School of Geoinformatics, and School of Biology, Institute of Science, Suranaree University of Technology. Sakaerat Environmental Research Station (SERS) for kindly facilitaing in field study. We kindly thank to Professor Dr. Dokrak Marod, Kasetsart University, for providing the AGB field data used in this study.

References

1. Lamb, W.F., Wiedmann, T., Pongratz, J., Andrew, R., Crippa, M., Olivier, J.G., et al.: A review of trends and drivers of greenhouse gas emissions by sector from 1990 to 2018. Environ. Res. Lett. **16**(7), 073005 (2021)
2. Yoro, K.O., Daramola, M.O.: CO$_2$ emission sources, greenhouse gases, and the global warming effect. In: Advances in carbon capture, pp. 3–28. Woodhead Publishing, Sawston (2020)

3. Alhorr, Y., Eliskandarani, E., Elsarrag, E.: Approaches to reducing carbon dioxide emissions in the built environment: low carbon cities. Int. J. Sustain. Built Environ. **3**(2), 167–178 (2014)
4. Freund, P.: Anthropogenic climate change and the role of CO^2 capture and storage (CCS). In: Geological Storage of Carbon Dioxide (CO_2), pp. 3–25. Woodhead Publishing, Sawston (2013)
5. Rodríguez-Veiga, P., Wheeler, J., Louis, V., Tansey, K., Balzter, H.: Quantifying forest biomass carbon stocks from space. Curr. For. Rep. **3**(1), 1–18 (2017)
6. Intergovernmental Panel on Climate Change (IPCC): IPCC guidelines for National Greenhouse Gas Inventories, IGES; Japan (2006)
7. Attarchi, S., Gloaguen, R.: Improving the estimation of above ground biomass using Dual Polarimetric PALSAR and ETM+ Data in the Hyrcanian Mountain Forest (Iran). Remote Sens. **6**(5), 3693–3715 (2014)
8. Attarchi, S., Gloaguen, R.: Carbon stock assessment using remote sensing and forest inventory data in Savannakhet, Lao PDR. Remote Sens. **6**(6), 5452–5479 (2014)
9. Ogino, K., Sabhasri, S., Shidei, T.: The estimation of the standing crop of the forest in north-eastern Thailand. Jpn. J. SE Asian Stud. **1**(4), 89–97 (1964)
10. Kanzaki, M., Kawaguchi, H., Kiyohara, S., Kajiwara, T., Kaneko, T., Ohta, S., ... Wachrinrat, C.: Long-term study on the carbon storage and dynamics in a tropical seasonal evergreen forest of Thailand. In: Tropical forestry change in a changing world, FORTROP II International Conference, pp. 17–20 (2008)
11. Ladpala, P., Sasrisang, A., Kaewpakasit, K.: Biomass and carbon stock of the dry dipterocarp forest, Sakaerat, Nakhon Ratchasima province. In: Proceedings of the Thailand Forest Ecological Research Network (T-FERN): ecological knowledge for adaptation on climate change, p. 135. Kasetsart University, Bangkok (2014)
12. Thongmee, T., Som-ard, J., Jitsukka, W.: Evaluation of above-ground carbon sequestration of forest in mahasarakham university using remote sensing data. J. Sci. Technol. MSU. **38**(6), 586–597 (2019)
13. Nguyen, L.D., Nguyen, C.T., Le, H.S., Tran, B.Q.: Mangrove mapping and above-ground biomass change detection using satellite images in coastal areas of Thai Binh Province, Vietnam. For. Soc. **3**(2), 248 (2019)
14. SERS homepage, https://www.tistr.or.th/sakaerat/sakaeratE/sakaeE.htm. Last accessed 25 Apr 2022
15. Sierra, C., del Valle, J., Orrego, S., Moreno, F., Harmon, M., Zapata, M., Colorado, Z.G., Herrera, M., Lara, W., Restrepo, D., Berrouet, L., Loaiza, L.: Total carbon stocks in a tropical forest landscape of the Porce region, Colombia. For. Ecol. Manag. **243**(2–3), 299–309 (2007)
16. Brown, S., Gillespie, A.J., Lugo, A.E.: Biomass estimation methods for tropical forests with applications to forest inventory data. For. Sci. **35**(4), 881–902 (1989)
17. Fuchs, H., Magdon, P., Kleinn, C., Flessa, H.: Estimating above-ground carbon in a catchment of the Siberian Forest tundra: combining satellite imagery and field inventory. Remote Sens. Environ. **113**(3), 518–531 (2009)
18. Liu, S., Cheng, F., Dong, S., Zhao, H., Hou, X., Wu, X.: Spatiotemporal dynamics of grassland above-ground biomass on the Qinghai-Tibet Plateau based on validated MODIS NDVI. Sci. Rep. **7**(1), 4182 (2017)
19. Wylie, B.K., Meyer, D.J., Tieszen, L.L., Mannel, S.: Satellite mapping of surface biophysical parameters at the biome scale over the North American grasslands. Remote Sens. Environ. **79**(2–3), 266–278 (2002)
20. Calvao, T., Palmeirim, J.M.: Mapping Mediterranean scrub with satellite imagery: biomass estimation and spectral behaviour. Int. J. Remote Sens. **25**(16), 3113–3126 (2004)
21. Marashi, M., Torkashvand, A.M., Ahmadi, A., Esfandyari, M.: The responses of Moso bamboo (Phyllostachys heterocycla var. pubescens) forest above-ground biomass to Landsat TM spectral reflectance and NDVI. Acta Ecol. Sin. **30**(5), 257–263 (2010)
22. Senpaseuth, P., Navanugraha, C., Pattanakiat, S.: The estimation of carbon storage in dry evergreen and dry dipterocarp forests in Sang Khom District, Nong Khai Province, Thailand. Environ. Natl. Resour. J. **7**(2), 1–11 (2009)
23. Marod, D.: Dynamic monitering of natural forest and forest restoration for sustainable management in the Sakaerat Environmental Research Station area (sub-research project) (2021) (In Thai)

Determination of Land Suitability for Oil Palm with Multi-dimension Decision Support Using Analytic Network Process (ANP) in Southern Thailand

Chomchanok Arunplod ⓘ, Apichon Witayangkurn ⓘ, and Daosaowaluk Kongtong ⓘ

Abstract Oil palm is one of the most important economic crops in Thailand. It contributes to the product in variety like petrol, cosmetics, consumer products, etc. The main problem for plantation oil palm plantation is improper land for plantation. Therefore, it results in fluctuations in reduction of quality and production quantity. The main objective of this study is to assess the land suitability for oil palm using multi-dimension criteria in Surat Thani and Krabi provinces, in Southern Thailand, through the analytic network process (ANP) integrated with geographic information system (GIS). The study was carried out with 28 layers grouped into six main criteria: topography, climate, physical soil, chemical soil, disaster events, and socioeconomic factors. All 28 criteria layers are weighted based on expert opinions according to the ANP method. The ANP disclosed that the rainfall is the most affecting criteria for oil palm plantation; the highest weight score is 0.112. Results from this study reveal two climate conditions: (1) land suitability for oil palm plantation based on 1-year climate data, a very highly suitable class is 273.6 km^2 (1.5%), and (2) oil palm suitability based on 5-year accumulated data, a very highly suitable class is 388.9 km^2 (2.2%). The suitable difference areas are affected by the variable rainfall and dry season conditions relating to the climate change situation at the local scale. Finally, this study developed an effective decision tool for local farmers to manage

C. Arunplod (✉)
Department of Geography, Faculty of Social Sciences, Srinakharinwirot University, Wattana, Bangkok, Thailand
e-mail: chomchanok@g.swu.ac.th

A. Witayangkurn
School of Information, Computer, and Communication Technology, Sirindhorn International Institute of Technology, Thammasat University, Khlong Luang, Pathum Thani, Thailand
e-mail: apichon@siit.tu.ac.th

D. Kongtong
Remote Sensing and GIS FoS, School of Engineering and Technology, Asian Institute of Technology, Klong Luang, Pathumthani, Thailand

© The Author(s), under exclusive license to Springer Nature Switzerland AG 2023
W. Boonpook et al. (eds.), *Applied Geography and Geoinformatics for Sustainable Development*, Springer Geography,
https://doi.org/10.1007/978-3-031-16217-6_17

217

their plantations with multi-dimensions to sustain the quality and quantity of oil palm yield.

Keywords Oil palm · ANP · GIS · MCDM · Multi-dimension criteria

1 Introduction

Oil palm is one of the most important crops for domestic consumption and export in Thailand [1]. It contributes to the product in variety like petrol, cosmetics, consumer products, etc. [2]. An equator country's zone is ideal for oil palm plantation; therefore, the global primary oil palm producers are Malaysia, Indonesia, and Thailand, accounting for more than 90% of the worldwide supply. Hence the quality of oil palm and the production should be high per area unit (rai) for suitable theoretical land [1]. Unlike the mono-crop plant in the neighboring countries, most oil palm plantations in Thailand are small- and medium-scale farms [2]. The oil palm plantations have been steadily expanding to meet the rising demand for biodiesel and food goods [3]. Several oil palm plantations take place on inappropriate lands, such as a previously shrimp farm, resulting in several adverse effects, such as reducing nutrients in the soil, which directly affect the quality of oil palm. Hence, the proper management due to the land conditions and crop requirements will increase the yield both in quality and quantity [4–6]. Previous researches focus on physical criteria or physical and climate consideration for oil palm plantation [6]. The challenging issue on suitability plantation is still there. In this study, the six main criteria of oil palm growing have been revealed, including topography, climate, physical soil, chemical soil, disaster, and socioeconomic factors.

The suitability analysis allows recognition of suitable regions, facilitating decision-makers to identify limiting factors and formulate appropriate management decisions to increase the productivity of the land. As aforementioned, this study discloses the land suitability for oil palm by considering multi-dimension criteria. There are 28 information layers collected from multiple sources grouped into the six main criteria.

One of the famous land suitability assessments is a Multi-Criteria Decision-Making (MCDM) model that considers several factors. Analytic network process (ANP) is one MCDM method that is adapted from the analytic hierarchy process (AHP) [7]. These two models were proposed by Saaty [8]. Furthermore, many variables and relationships can be broken down into manageable information units using ANP [9].

A geographic information systems (GIS) has become influential and respected by decision-makers, associated with MCDM, and has been widely adopted in environmental planning, engineering, agriculture plantation, etc. [7, 9]. GIS is a helpful tool for storing, managing, and analyzing [10]. In addition, the surface and overlay analysis capabilities in GIS can effectively facilitate handling vast amounts of spatial information. The powerful query, analysis, and integration mechanism of GIS make an ideal scientific tool to analyze land suitability.

All suitability criteria have been emphasized. Therefore, the importance scales, called weight scores, were assigned to each of these 28 layers based on the opinion of experts. ANP tool is used to compare complex data, manage a set of data, and use expert opinion to make the best selection among the relating variables.

This study aims to determine land suitability for oil palm plantations, with multi-dimension criteria associated with GIS and ANP to overcome the classical criteria analysis. The primary output is suitability maps of oil palm with two climate conditions which publicize the climate change impact on local scale plantations. The critical finding of this research utilizes the socioeconomic factors for land suitability to represent the dynamic phenomena of oil palm plantations.

1.1   Land Characteristics for Oil Palm

Many studies mentioned that land characteristics are considered when determining land suitability for sustainable oil palm. The work of Paramananthan [11] explains land characteristics suitable for oil palm as several physical parameters such as climate, topography, wetness, and physical and chemical soil conditions. A subcategory of land characteristics can be found within each of these five groups [11]. The details of suitable oil palm plantations are as follows: Climatic factors used in the land evaluation process include dry season, annual rainfall, annual temperature, annual maximum temperature, and annual minimum temperature. Topography or slope has an impact on soil erodibility. Drainage and flooding are defined as wetness. Soil texture/structure and soil depth are physical soil, etc. [11]. Those physical criteria are considered as same as Kome et al. [12] who studied land requirements for the production of oil palm. Land requirements for the production of oil palm have four main criteria including climate factors, topography, physical properties, and soil fertility. Kome et al. [12] focused on the soil fertility conditions defined as cation exchange capacity (CEC), pH at the topsoil, base saturation, organic carbon mg:k ratio, and K (mole fraction). Salinity and alkalinity are electrical conductivity (EC) [12]. Moreover, the palm oil biophysical suitability assessment studies three main criteria. Other criteria are proposed by researchers including soil properties like soil texture, depth, and coarse fragments. The topography includes elevation and slope [13, 14]. An important of soil characteristics for oil palm including peat thickness, pear ripening and water source including groundwater, and distance from the river criteria to determine the land suitability assessment for oil palm production [13].

Not only are the criteria related to land suitability, but the essential factors for oil palm growth are also a consideration. The climate suitability index for oil palm includes solar radiation, rainfall, dry season, mean annual temperature, and wind. Topography is the slope. Drainage includes drainage class and flooding [15, 16]. Physical soil conditions have texture, depth sulfate layer, root restricting layer, and thickness of organic soil. Chemical soil conditions include effective CEC, base saturation in A horizon, organic C in A horizon, salinity 50 cm depth, and micronutrients [15]. In addition, chemical soil includes pH, organic matter, the salinity of the soil, nitrogen, phosphorus, potassium, magnesium, and CEC [16].

1.2 Land Suitability Analysis in GIS

Land suitability assessment is a systematic method capable of determining the suitable location for specific purposes [4], such as business area, human settlement, aquaculture, and agriculture using geographic information system (GIS) technology [17]. Several research studies have been undertaken to assess the land suitability assessment for agriculture based on integrated techniques using GIS, including back-propagation neural network (BPNN) [18], fuzzy set theory [19], decision-making trial and evaluation laboratory (DEMATEL) [20], simple additive weighting (SAW) [21], artificial neural network (ANN) [22], machine learning (ML) [23], analytic hierarchy process (AHP) [10, 20, 24, 25], and analytic network process (ANP) [19, 26].

Oil palm farmers can use the combination of multi-criteria decision analysis (MCDA) and GIS to identify the most suitable areas for oil palm production and determine the future direction of oil palm production [13]. In Southern Thailand, studies have developed a methodology for land evaluation for oil palm plantations using AHP evaluation in GIS [27].

This study combines the multi-criteria decision-making based on the geographic information system (GIS-MCDM) approach and the analytic network process (ANP) [28]. An essential feature of the proposed multi-criteria model is that the ANP considers the interdependence between different criteria. ANP and GIS-MCDM potential areas were determined based on experts' opinions on the most important or limiting factors. However, few studies have attempted to develop a GIS combined with ANP analysis for MCDM to determine sustainable crop production for future planning [26].

2 Study Area

The study is located in Surat Thani and Krabi provinces, the Southern part of Thailand with an approximate area of 18,030.4 square kilometers (km²). Geographically, Surat Thani and Krabi provinces are located between 7°47′ and 9°80′ north latitude and 98°43′ and 100°09′ east longitude (see Fig. 1). Due to the influence of the tropical monsoon, Surat Thani and Krabi provinces experience only two seasons: the summer season (January–April) and the rainy season (May–December). In the part, this area is well known for the great rubber plantation and its role as a major para-rubber exporter. However, the change in agriculture policies

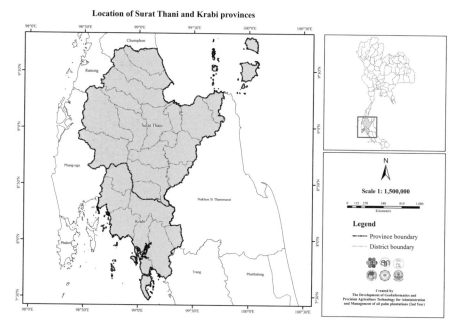

Fig. 1 Surat Thani and Krabi province, Thailand

in the last decade affected land-use change in this area. The agricultural land turns into coconut and oil palm plantations, and presently the major crops of Surat Thani and Krabi provinces are oil palm and rubber. In 2019, the million rai of oil palm project is promoted, increasing the demand for oil palm plantation in this area.

3 Methodology

The land suitability for oil palm is investigated in Surat Thani and Kribi provinces using the ANP framework associated with the GIS-overlay technique. The overall methodology is divided into three parts: (1) data collection, (2) weight score analysis using ANP, and (3) GIS processing as demonstrated in Fig. 2.

3.1 Data Collection

Data collection was obtained from various sources and map scales, due to the data available. Based on the literature criteria [6, 29, 30], the data collected were aggregated for topography, physical soil, chemical soil, and disasters which infer to the environmental situation, climate, and additional socioeconomic were considered for

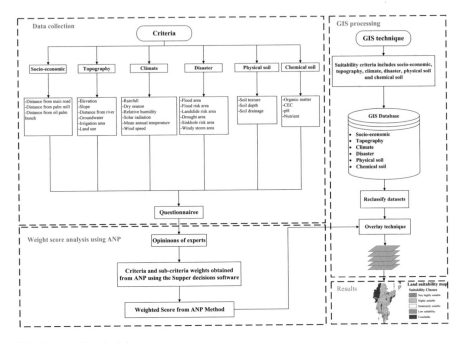

Fig. 2 Overall methodology

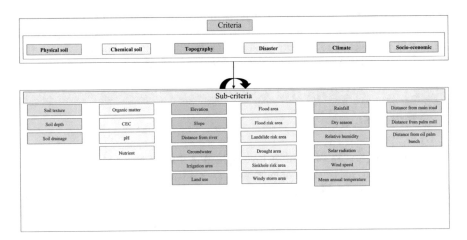

Fig. 3 Criteria and sub-criteria for oil palm

the oil palm plantations. The 28 variables can be subdivided into six groups (see Fig. 3):

(i) Topography included elevation, slope, distance from the river, irrigation area, groundwater, and land use.

(ii) Physical soil had soil texture, soil depth, and soil drainage.

(iii) Chemical soil included pH, organic matter, CEC, and nutrient.
(iv) Climate included mean temperature, relative humidity, wind speed, rainfall, dry season, and solar radiation.
 (v) Disaster included flood area, flood risk area, landslide risk area, drought area, sinkhole risk area, and windy storm area.
(vi) Socioeconomic included distance from the main road, distance from palm mill, and distance from oil palm bunch.

The aforementioned criteria were collected from field surveys and secondary data. Field survey data was collected from oil palm bunch, palm mill, and land use (LU) using the GPS and data logger. The secondary data was acquired from the government and open sources: climate data, soil properties, disaster data, land use, groundwater, irrigation area, road, and national park. Then all criteria must be verified or weighted by oil palm expertise; therefore, the questionnaire is designed and evaluated. The challenge at this point is the data variety. The collected data are various in source, map scale, data type, and an acquiring date as Table 1. All data is set to transform into a ready format for data analysis using the Super Decisions software version 2.2

3.2 Weight Score Analysis Using ANP

The different aspects of all criteria were explained based on analytic network process (ANP) in which criteria are grouped into clusters of related factors rather than into hierarchical levels [28]. For that reason, the expert's opinions are required to score the importance of the criteria [27, 28]. Experts from a variety of oil palm-related sectors were interviewed. Therefore, there are 19 experts who advised and weighted criteria for this study, including the six experts from the Office of Agricultural Research and Development Region 7, four experts from Surat Thani Oil Palm Research Center, five oil palm farmers in Surat Thani and Krabi to collect expert information on oil palm plantation issues, and four GIS experts from another sector as the technical method suggestor. There are 28 criteria that were selected for assessing land suitability for oil palm (see Table 2), based on an extensive literature review of land suitability for oil palm and the recommendations from the Office of Agricultural Research and Development Region 7.

In this study, 28 criteria that aid in decision-making are measured by using the ANP model (see Fig. 4). Weights were allocated for each criterion using the ANP introduced by Saaty. In the ANP, individual preferences or judgments are expressed using a fundamental measurement of 9 points, which is used to create a matrix of pairwise comparisons.

The pairwise comparisons were applied for both critical criteria and sub-criteria. Each criterion is compared with others using AMP comparisons as Table 3, where a value of 1 means equal importance, and a value of 9 is extreme significance over another factor, respectively. Accordingly, weights were assigned relevant to criteria

Table 1 Characters and source of data

Criteria	Type of data	Year	Scale	Source
Experts' opinion	Text	2020		Questionnaire
Windy storm area	Text	2018–2020		Department of Disaster Prevention and Mitigation
Climate data	Text	2016–2020		Thai Meteorological Department
Palm mill	Point	2020		Filed survey
Oil palm bunch	Point	2020		Filed survey
Soil properties	Vector	2018	1:25,000	Land Development Department
Land use	Vector	2018	1:25,000	Land Development Department
Flooding area	Vector	2014–2018	1:50,000	Land Development Department
Landslide risk area	Vector	2005–2016	1:50,000	Land Development Department
Drought area	Vector	2014–2018	1:50,000	Land Development Department
Groundwater	Vector		1:100,000	Department of Groundwater Resources
Sinkhole risk area	Vector	2020	1:250,000	Digital Government Development Agency[a]
Irrigation area	Vector		1:25,000	Regional Irrigation Office 15
Road	Vector		1:20,000	Royal Irrigation Department[b]
River	Vector	2019		Department of Water Resources[c]
Flood risk area	Vector	2020		Department of Water Resources[d]
National park	Vector			Department of National Parks, Wildlife and Plant Conservation[e]
Solar radiation	Raster	2016–2020	50 km	NASA Power[f]
Digital Elevation Model	Raster	2009–2012	12.5 m	Advanced Land Observing Satellite (ALOS)

[a]DGA: https://data.go.th/dataset/sinkhole
[b]RID: http://water.rid.go.th/hydhome/hydrology/irr_area.php
[c]DWR: http://division.dwr.go.th/DOC/index.php/fgds
[d]DWR: http://mekhala.dwr.go.th/download.php
[e]DNP: http://www2.dnp.go.th/gis/
[f]NASA: https://power.larc.nasa.gov/data-access-viewer/

based on expert ideas and literature [19, 25]. The pairwise comparison is pre-evaluated and retrieved from the Super Decision Software version 2.2, which calculates the relative weights of the various elements of the matrices and the consistency ratio (CR) employed for locating the inconsistent ratings.

The weight for each factor was calculated through a pairwise comparison matrix, and the maximum eigenvalues (λ_{max}) of the normalized matrix were computed. The consistency index (CI) was estimated using the formula: $CI = (\lambda_{max} - n)/(n - 1)$. The random consistency index (RCI) was used to determine the degree of consistency or consistency ratio (CR) (i.e., CI/RCI). If the CR value is less than or equal to 0.1, the inconsistency is acceptable, or the pairwise comparison may be revised.

Table 2 Evaluation of land characteristics for oil palm

Criteria	Range of suitability					References
	Very highly suitable (5)	Highly suitable (4)	Moderately suitable (3)	Low suitability (2)	Unsuitable (1)	
Topography						
Elevation (m)	0–400	–	400–500	500–900	>900	[12]
Slope (%)	0–4	4–12	12–23	23–38	>38	[15]
Distance from river (m)	<500	–	500–1000	1000–1500	>1500	[12]
Groundwater	High	–	Moderate	Low	Lack	[12]
Irrigation area	Irrigation area	–	–	–	Not irrigation area	
Land use	Cropland and rangeland	–	–	–	Urban and built-up land, roads, waterbody, miscellaneous land, national park	
Physical soil						
Soil texture	SL, L, SiL	CL, SiCL, SC	SCL, LS, SiC	C	Gravel, S	[15]
Soil depth (cm)	>100	75–100	50–75	25–50	<25	[15]
Soil drainage	Moderate	Good to excessive	Excessive	Very poor	Very excessive	[15]
Chemical soil						
Organic matter	>2.5	2.5	1.5	1.2	<0.8	[15]
CEC (meq/100g)	>18.0	18.0	15.0	12.0	<6.0	[15]
pH	>5.5	5.5	4.2	4.0	<3.5	[15]
Nutrient	High	–	Moderate	–	Low	
Climate						
Rainfall (mm/year)	2500–3500	1700–2500	1450–1700	1250–1450	<1250	[15]
Dry season (<100 mm/month)	None	1	1–2	2–3	>3	[14]
Relative humidity (%)	>75	70–75	60–70	-	<60	[11]
Solar radiation (MJ/m²)	13–15	11–13 15–17	20–22 17–19	7–9 19–21	>21	[15]

(continued)

Table 2 (continued)

Criteria	Range of suitability					References
	Very highly suitable (5)	Highly suitable (4)	Moderately suitable (3)	Low suitability (2)	Unsuitable (1)	
Mean annual temperature (°C)	25–29	22–25 29–32	20–22 32–35	16–20 35–37	<16 >37	[15]
Wind speed (m/s^{-1})	5–8	3–5 8–10	<3	15–20 >10–15	>20	[15]
Disaster						
Flood area	Not flooded	–	Minor	Moderate	Severe	[27]
Flood risk area	Not flood risk	–	Minor	Moderate	Severe	[27]
Landslide risk area	Not landslide	–	Minor	Moderate	Severe	[27]
Drought area	Not drought	–	Minor	Moderate	Severe	
Sinkhole risk area	Not sinkhole	–	Minor	Moderate	Severe	
Windy storm area	Not windy storm	–	Minor	Moderate	Severe	
Socioeconomic						
Distance from main road (km)	<2	2–3	3–4	4–5	>5	
Distance from palm mill (km)	<20	20–40	40–80	80–120	>120	
Distance from oil palm bunch (km)	<5	5–10	10–15	15–20	>20	

According to oil palm experts, for the application of the ANP, the criteria were weighted and scored based on their significance for oil palm plantation. The Super Decisions software was used to confirm the final weight score for all criteria for ANP as Table 4, which defined the most significant criteria for oil palm plantations as rainfall (0.112). The second importance is the soil texture (0.111), soil drainage (0.068), dry season (0.064), and nutrient (0.055), respectively. On the other hand, the most negligible impact on land suitability evaluation for oil palm was the landslide risk area and windy storm area, accounting for 0.004.

For instantaneous consideration, the weight of critical criteria shown as climate (0.246) is most significant for oil palm plantation, followed by physical soil (0.218), topography (0.202), chemical soil (0.191), socioeconomic (0.087), and disaster (0.056), respectively (see Fig. 5).

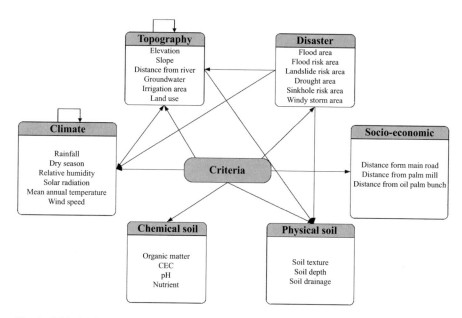

Fig. 4 Critical criteria in ANP framework in Super Decisions software

Table 3 Scales for the pairwise ANP comparisons [7]

Intensity of importance	Definition	Explanation
1	Equal importance	Two activities contribute equally to the objective
2	Weak or slight	
3	Moderate importance	Experience and judgment slightly favor one activity over another
4	Moderate plus	
5	Strong importance	Experience and judgment strongly favor one activity over another
6	Strong plus	
7	Very strong or demonstrated importance	An activity is favored very strongly over another; its dominance demonstrated in practice
8	Very, very strong	
9	Extreme importance	The evidence favoring one activity over another is of the highest possible order of affirmation
2,4,6,8	Intermediate value	Used to represent a compromise between the priorities listed above

Table 4 The final weight score for sub-criteria derived from ANP procedure

Rank	Criteria	Weight
1	Rainfall	0.112
2	Soil texture	0.111
3	Soil drainage	0.068
4	Dry season	0.064
5	Nutrient	0.055
6	Distance from river	0.053
7	pH	0.047
8	Soil depth and organic matter	0.042
10	Elevation	0.041
11	Irrigation area	0.038
12	Land use	0.037
13	Slope	0.035
14	Solar radiation	0.034
15	Distance from palm bush	0.032
16	Mean temperature	0.031
17	Relative humidity	0.028
18	Groundwater	0.027
19	Distance from palm mill	0.026
20	CEC	0.014
21	Distance from main road and flooding area	0.013
23	Drought area	0.009
24	Wind speed and flood risk area	0.007
26	Sinkhole risk area	0.006
27	Landslide risk area and windy storm area	0.04

Fig. 5 Weight score of criteria

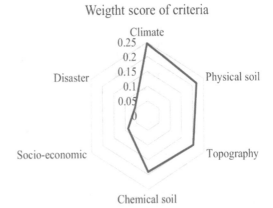

Weigtht score of criteria

3.3 GIS Processing

All weighted criteria are prepared in GIS-ready format, and the geographic information system (GIS) software named ArcGIS 10.8 is selected to generate a suitability map by the following:

Step 1: Identify the assessment criteria.

For each criteria, we identified and determined the value of suitability classes for oil palm growth, ranked 1–5 according to the range of suitability in Table 2.

Step 2: Involve preparing the layers and workspace to generate and ensure all layers were set up in the same coordinate system and reference projection.
Step 3: Involve preparing data.

Climate data include rainfall, mean annual temperature, solar radiation, relative humidity, dry season (the number of months that receive less than 100 mm of rainfall), and wind speed. The climate data were interpolating through the inverse distance weighted (IDW) method to generate raster images into 12.5 m from meteorological stations distributed throughout Surat Thani and Krabi provinces.

Socioeconomic data includes distance from oil palm bunch, palm mill, and main road. Oil palm bunch and palm mill were collected in vector (point) format, and main road was obtained in vector (line) format. The distance between oil palm bunches, palm mills, and main roads was calculated using the Euclidean distance tool and transformed into 12.5 m raster data.

Topography data includes elevation, slope, and distance from river. Digital Elevation Model calculates both elevation and slope. Distance from the river was generated by Euclidean distance tool and transformed into 12.5 m raster data.

Step 4: Involve reclassifying all raster to a standard suitability scale using the suitability thresholds identified from Step 1, including climate data, socioeconomic, elevation, slope, and distance from river layers. Factors are given numerical values that indicate if their presence is not suitable for the growth of palm oil or is favorable to the development of palm oil. Next, reclassify each dataset to a standard scale (1–5), giving higher values to more suitable attributes.
Step 5: Involve reclassifying all vectors to a standard suitability scale using the suitability thresholds identified from Step 1, including physical soil, chemical soil, disaster, land use, irrigation area, and groundwater layers. Next, reclassify each dataset to a standard scale (1–5), giving higher values to more suitable attributes.
Step 6: Require data conversion polygon to raster data. Data were obtained in vector format and were transformed into 12.5 m raster data.
Step 7: Perform an overlay analysis with all the selected variables to create a suitability map, reclassify the suitability results, and convert to the desired suitability scale and projection system. In this step, a land suitability map for oil palm was created using the weighted overlay tool by assigning a given weight to each criterion. Following the average weights of the ANP analysis, each raster was

assigned a weight. Next, each raster value was multiplied by its weight and then added to create the output raster.

The formula used to calculate the land suitability index is as follows (Eq. [19]):

$$S_i = \sum X_i W_i \qquad (1)$$

where, X_i = value of each criterion

W_i = weight values of each criterion
S_i = suitability index

The total suitability scores range between 0 and 5 and are divided into five suitability classes, i.e., very highly suitable (4–5), highly suitable (3–4), moderately suitable (2–3), low suitability (1–2), and unsuitable (0–1).

Step 8 removed exclusionary areas (i.e., urban, roads, water bodies, rivers, national parks).

Consequently, the final results are suitability maps that represent areas from suitable to not suitable for the oil palm plantation.

The validation method concerned whether the selected crop had already been produced or grown in the region. The final suitability map produced from the model was verified to ensure that the model corresponded to the actual conditions in the field. Then, using the geometry intersection tool in ArcGIS software, each suitability class was intersected with the existing oil palm area layer.

4 Land Suitability Map for Oil Palm Plantation

All criteria have been standardized (reclassified) into five classes (very highly suitable, highly suitable, moderately suitable, low suitability, and unsuitable). This study used two intervals of annual climate data, 1 year (2020) and 5 years (2016–2020), to represent the land suitability for oil palm plantation due to climate change effect.

4.1 Land Suitability Map for Oil Palm Using Annual Climate Data for Year 2020

Analysis of suitability for oil palm assessment produced with annual climate data for year 2020 shows that the 1.5% (273.6 km²) of the total land in study areas is defined as very suitable for oil palm plantation, 60.7% (10,946.5 km²) is highly suitable, 6.1% (1102.5 km²) is moderately suitable, 3.3% (591.3 km²) is low suitability, and 28.4% of the study area (5116.5 km²) is unsuitable (see Fig. 6). The unsuitable

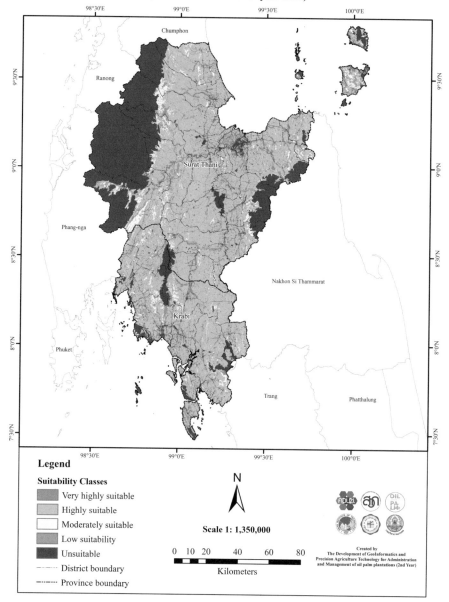

Fig. 6 Land suitability for oil palm using annual climate data for the year 2020

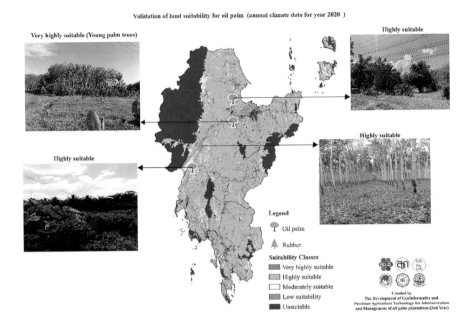

Fig. 7 Validation of land suitability for oil palm using annual climate data for the year 2020

area is where oil palm could not be grown (i.e., urban, roads, water bodies, rivers, national parks). Therefore, most of the land suitability assessments for oil palm are highly suitable and very highly suitable classes with existing oil palm growing areas. The land suitability map was validated using an on-path survey (see Fig. 7), where we found that the existing young oil palm plantation is in a very highly suitable area. In addition, some existing oil palm plantation is in highly suitable area, and some young oil palm has been replaced in highly suitable area.

4.2   Land Suitability Map for Oil Palm Using Average Annual Climate Data from 2016 to 2020

Figure 8 retrieves the land suitability map for oil palm plantation using annual climate data from 2016 to 2020 in each suitability class. The land consists of an estimate of 2.2% (388.9 km²), 60.2% (10,846.7 km²), 6.7% (1218.7 km), 2.5% (459.6 km²), and 28.4% (5116.5 km²) of the study area as very highly suitable, highly suitable, moderately suitable, and low suitability, respectively. Therefore, most of the land suitability assessments for oil palm are highly suitable and very highly suitable classes with existing oil palm growing areas (see Fig. 9).

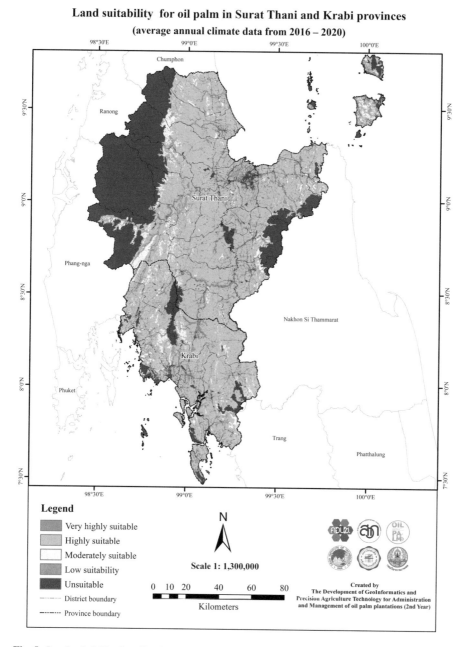

Fig. 8 Land suitability for oil palm using average annual climate data from 2016 to 2020

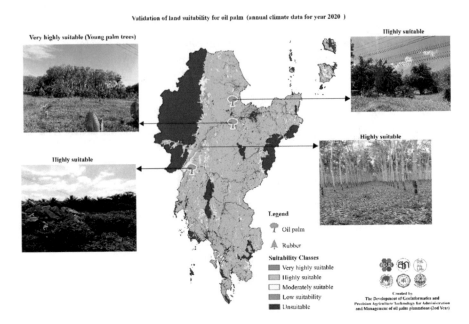

Fig. 9 Validation of land suitability for oil palm using average annual climate data from 2016 to 2020

5 Conclusion and Discussion

The land suitability analysis for oil palm plantation is an essential information not only for the developer and oil palm farmer but also the governors or land planners. This study used GIS processing and analysis in conjunction with the ANP technique to assess land suitability for oil palm plantation. The spatial land suitability for oil palm was obtained after analyzing selected evaluation criteria such as climate, topography, physical soil, chemical soil, disasters, and socioeconomic factors. As a result, the land suitability assessment for oil palm using annual climate data of 1 year (2020) falls into very highly suitable 1.5%, highly suitable 60.7%, moderately suitable 6.1%, low suitability 3.3%, and unsuitable 28.4%. Land suitability using annual climate data from 2016 to 2020 gave an estimate of 2.2%, highly suitable 60.2%, moderately suitable 6.7%, low suitability 2.5%, and unsuitable 28.4%. Climate factors have an impact on land suitability analysis for oil palm. The land suitability assessment utilized in this study provides a helpful decision-making approach to help oil palm growers and decision-makers carry out suitability assessments for oil palm. Notably, socioeconomic and disaster factors allow farmers to make better decisions in their farming operations.

The integrated approach of GIS and the ANP technique with socioeconomic criteria in this study has a great potential to classify the land suitability of oil palm plantations. No previous studies have been conducted in Southeast Asia. This holistic multi-dimension data approach of oil palm suitability assessment for the first time, which the physical data and socioeconomic data were integrated.

Previous land suitability assessments only assessed Thailand's climate, topography, and soil for oil palm growth and production [2, 26]. The multi-dimension criteria were divided into six groups: topography, physical soil, chemical soil, disaster, climate, and socioeconomic to assess land suitability for oil palm plantation. These criteria are all maps of criteria; and the maps were the same spatial coordinate reference to the same scale, boundary extent, resolution, and spatial reference before they were overlay technique.

Based on the questionnaire and its assigned weight, rainfall got the highest weight (0.112) among all criteria in this study (see Table 4). Soil texture is the second critical rank factor in the study area, with a weight of 0.111. With a weight of 0.068, soil drainage is third essential. Dry season (0.064) and nutrient (0.055) are fourth and five rank factors, respectively. Rainfall and dry season are the most factors that impact oil palm growth because water affects nutrient movement [29]. In addition, soil properties have involved oil palm production [14].

The climate is the most critical factor that determines the land suitability of oil palm. This study used two intervals of annual climate data: 1 year (2020) and 5 years (2016–2020). The results show that using average annual climate data for 5 years has land suitability more than climate data for a single year. The land suitability assessment for oil palm using annual climate data 1 year (2020) falls into very highly suitable 273.6 km^2 (1.5%) and land suitability using average annual climate data from 2016 to 2020 falls into very highly suitable and gave an estimate of 388.9 km^2 (2.2%).

According to land characteristics for oil palm, we found that the number of months that receive less than 100 mm of rainfall (dry season) more than 3 months per year of climate data 1 year is unsuitable for oil palm. Because climate data in a single year is warmer or colder, it is not enough to confirm that the climate has changed.

Rainfall is the most important for determining the suitability of areas for oil palm plantation, associated with the dry season, which causes water deficiency to impact oil palm growth. In addition, climate change affects oil palm growth and products. Tropical countries in particular are affected negatively by climate change which also has a detrimental impact on oil palm agronomy [29, 31].

The results from this study reveal that the existing oil palm growing areas represent highly and highly appropriate classes, implying robust modeling and appropriate weights. Furthermore, the high overlays with current oil palm growing lands in the very high and highly suitable class prove the selection methods of farmers and demonstrate the validity of the methodology applied here.

The land suitability assessment for oil palm plantations in the maps (see in Figs. 6 and 8) can be used to identify the suitable area for oil palm. Farmer can also improve the suitability of available lands for oil palm plantation by adopting appropriate agronomy measures relevant to limited factors indicated. This study has provided information for the oil palm plantation sector and helps the farmer to easily identify the suitable area for oil palm plantation and the limiting factors.

References

1. Agricultural Research Development Agency (ARDA). Environmental Factors Affecting Plant Growth., https://www.arda.or.th/kasetinfo/south/palm/controller/01-03.php. Last accessed 18 Dec 2020
2. BangkokPost. Making palm oil more sustainable., https://www.bangkokpost.com/business/2048875/making-palm-oil-more-sustainable. Last accessed 15 July 2021
3. Jaroenkietkajorn, U., Gheewala, S.H.: Land suitability assessment for oil palm plantations in Thailand. Sustainable Production and Consumption. **28**, 1104–1113 (2021)
4. Driessen, P.M., Konijn, N.T.: Land-Use Systems Analysis. WAU and Interdisciplinary Research (INRES). The Netherlands (1992)
5. Everest, T., Sungur, A., Özcan, H.: Determination of agricultural land suitability with a multiple-criteria decision-making method in Northwestern Turkey. International Journal of Environmental Science and Technology. **18**(5), 1073–1088 (2021)
6. Jantaraniyom, T., et al.: Oil Palm Plantations Management Effects. Prince of Songkhla University, Songkhla (2007)
7. Azizi, A., et al.: Land suitability assessment for wind power plant site selection using ANP-DEMATEL in a GIS environment: case study of Ardabil province, Iran. Environmental monitoring and assessment, 186 (2014)
8. Thomas L. Saaty, Vargas, L.G.: Decision making with the analytic network process. 2 ed. International Series in Operations Research & Management Science, Springer, Boston, MA. XVII, 363 (2013)
9. Eftekhari, E., Mahdavi, M.: Land suitability assessment using ANP in a GIS environment for Tourism Development Site (Case study: Lavasan-e Kuchak Rural District, Tehran province, Iran). Iranian Journal of Tourism and Hospitality Research. **7**(1), 5–17 (2019)
10. ElSheikh, R.F., et al.: An agricultural investment map based on geographic information system and multi-criteria method. Journal of Applied Sciences. **10** (2010)
11. Paramananthan, S.: Soil suitability and management implications of soil taxonomy with special reference to tree crop cultivation (1987)
12. Kome, G., et al.: Land suitability evaluation for oil palm (Elaeis guineensis Jacq.) in Coastal Plains of Southwest Cameroon. Open. Journal of Soil Science. **10**, 257–273 (2020)
13. Johaerudin and Nakagoshi, N.: GIS-based land suitability assessment for oil palm production in Landak Regency, West Kalimantan. Hikobia, 16, pp. 21–31 (2011)
14. Hashemvand Khiabani, P., Takeuchi, W.: Palm oil bio-physical suitability assessment in Indonesia and Malaysia over 2003–2015 (2018)
15. Nilnond, C., Jantaraniyom, T.: Oil Palm Plantations Management Effects. In: Thailand Research Expo 2008, p. 37 (2008)
16. Office of Agricultural Research and Development Region 7, Improving the efficiency of oil palm production (2011)
17. Stoms, D.M., McDonald, J.M., Davis, F.W.: Fuzzy assessment of Land suitability for scientific research reserves. Environmental Management. **29**(4), 545–558 (2002)
18. Kong, C., et al.: Geo-environmental suitability assessment for agricultural land in the rural–urban fringe using BPNN and GIS: a case study of Hangzhou. Environmental Earth Sciences. **75**(15), 1136 (2016)
19. Seyedmohammadi, J., et al.: Integration of ANP and fuzzy set techniques for land suitability assessment based on remote sensing and GIS for irrigated maize cultivation. Archives of Agronomy and Soil Science. **65**(8), 1063–1079 (2019)
20. Layomi Jayasinghe, S., Kumar, L., Sandamali, J.: Assessment of potential land suitability for Tea (Camellia sinensis (L.) O. Kuntze) in Sri Lanka using a GIS-based multi-criteria approach. Agriculture. **9**(7) (2019)
21. Mendas, A., et al.: Improvement of land suitability assessment for agriculture—application in Algeria. Arabian Journal of Geosciences. **7**, 435 (2013)

22. Ahmadi, F., Farsad Layegh, N.: Integration of artificial neural network and geographical information system for intelligent assessment of land suitability for the cultivation of a selected crop. Neural Computing and Applications. **26**, 1311–1320 (2015)
23. Taghizadeh-Mehrjardi, R., et al.: Land suitability assessment and agricultural production sustainability using machine learning models. Agronomy. **10**(4) (2020)
24. Ahmed, G.B., et al.: Agriculture land suitability analysis evaluation based multi criteria and GIS approach. IOP Conference Series: Earth and Environmental Science. **37**, 012044 (2016)
25. Akıncı, H., Özalp, A.Y., Turgut, B.: Agricultural land use suitability analysis using GIS and AHP technique. Computers and Electronics in Agriculture. **97**, 71–82 (2013)
26. Zabihi, H., et al.: Land suitability procedure for sustainable citrus planning using the application of the analytical network process approach and GIS. Computers and Electronics in Agriculture. **117**, 114–126 (2015)
27. Dansagoonpon, S., Tripathi, N.K.: Modeling site suitability for oil palm plantations in Southern Thailand. GIScience & Remote Sensing. **43**(3), 252–267 (2006)
28. Creative Decisions Foundation – CDF homepage, https://www.superdecisions.com/method/. Last accessed 21 Nov 2021
29. Hemthanon, U.K., Kaveeta, R.: Land suitability evaluation for economic crops in Khuan Kreng Peatland. Journal of Southern Technology. **10**, 135–145 (2017)
30. Jantaraniyom, T.: Ecology for Oil Palm. Prince of Songkhla University, Songkhla (2010)
31. Paterson, R.R.M., Lima, N.: Climate change affecting oil palm agronomy, and oil palm cultivation increasing climate change, require amelioration. Ecology and Evolution. **8**(1), 452–461 (2017)

Land Use Change and Ecosystem Service Variations in Huai Luang River Basin, Udon Thani Province, Thailand

Sathaporn Monprapussorn

Abstract Huai Luang river basin is one of the sub-basins of Mekong River that supports socioeconomic development in northeast of Thailand. Alteration of natural ecosystems into built-up and agricultural land often causes damage to ecosystem services in a basin. This chapter aims at projecting land use change in 2017–2032 by considering six biophysical and socioeconomic drivers: soil group, distance from road, rail and river, population density, and rice suitability area. Two land use scenarios, business as usual (BAU) and ecosystem service (ES), have been used to analyze feedbacks to ecological system. Results reveal the increase in built-up land for BAU scenarios, while wetland and grassland tend to be increased for ES scenarios. Ecosystem service value of ES scenario in 2032 increases by 26.19% compared to those of BAU scenario. Results are useful for future ecosystem monitoring, land use planning, and spatial decision-making at a basin and/or provincial level to promote sustainable land use planning.

Keywords Land use · Ecosystem services · Udon Thani

1 Introduction

Ecosystem provides resources as goods and services to support human well-being while simultaneously maintaining ecosystem functions and biodiversity. An advancement in production technology had led to increase in yields, timber, housing and other commercial commodity at the cost of ecosystem service reduction [3]. Rapid urban expansion in developing countries has placed the great pressures that could interfere with an ecosystem's ability to provide the services [4]. Changing in

S. Monprapussorn (✉)
Department of Geography, Faculty of Social Sciences, Srinakharinwirot University, Bangkok, Thailand
e-mail: sathaporn@g.swu.ac.th

© The Author(s), under exclusive license to Springer Nature Switzerland AG 2023
W. Boonpook et al. (eds.), *Applied Geography and Geoinformatics for Sustainable Development*, Springer Geography,
https://doi.org/10.1007/978-3-031-16217-6_18

land use can increase the risk of ecosystem degradation and human well-being and is expected to continue at heightened risks in the future [5]. Land use is often changed by different drivers, i.e., crop patterns, fluctuation in a crop prices, urbanization, agroindustry promotion, and food security. There are two main drivers for land use change: biophysical and socioeconomic drivers [6]. Biophysical drivers involve many characteristics of physical and biological factors, those that spend long period of time to change, for example, soil type, geology, and distance from transportation routes. Socioeconomic drivers are changed according to policy issues, i.e., population density, land policy, and crop prices. There are even greater uncertainties with respect to socioeconomic drivers in emerging economies, given the complex and unpredictable interactions between socioeconomic factors and land use.

Scenario-based assessment is widely used for simulating future land use by exploring past interaction and linkage between human and surrounding environment to predict future development pathways [7]. Land use scenarios can be expressed in two ways: quantitative and qualitative scenarios [1]. The advantage of qualitative scenario is the capability of adding several land use perspectives based on different stakeholders without using technical specification, leading to filling the gap among experts, decision-makers, and stakeholders [8]. The development of land use model should focus on global, regional, and local scales and consider driving forces, processes, and impacts on future scenarios [9]. Land use projection can act as a proxy by coordinating the land cover type proportional to each predefined land use class and assign the economic values, including value of coefficient set [10]. This chapter focuses on Huai Luang river basin, one of the sub-basins of Mekong River, which contribute significantly to economic growth in northeast region of Thailand. The basin extends over three provinces and most of the basin areas lie within Udon Thani province. A basin also provides many essential resources for local livelihoods, i.e., water, food, fishery, and agriculture. However, land use has been changed from miscellaneous land to urban area in recent decades, which then affects the ecosystem services. It is therefore important to assess ecosystem service value in comparison with the simulated ecosystem service value under different land use scenarios in Huai Luang river basin between 2017 and 2032. Results can be useful to provide guidance for planners and decision-makers to improve ecosystem services and increase resilience through land use planning.

2   Methodology

Land use assumption is developed to provide the basis for growth projections for Huai Luang river basin in 2032. The analysis of secondary data involves reviewing reports, policy, and strategic plans for Huai Luang river basin to extract some key driving forces for land use change and pathways for river basin development.

Business as usual (BAU) and green growth (GG) scenarios are constructed at basin level by retrieving essential information, literatures, and related documents. Land use has been grouped into seven classes: cultivated land, forestland, grassland, wetland, unused land, construction land, and water body.

Six physical and socioeconomic drivers consist of distance to road, river and rail, population density, soil groups, and rice suitability area. The conversion resistance of land use is one of the specific settings to determine the temporal dynamics of the simulation. Land use type with high capital investment cannot be easily converted to other uses such as urban and natural water. Therefore, dimensionless factors have been assigned to each land use, ranging from 0 (easily conversion) to 1 (irreversible change). Land use services define functional role of each land use and land use with similar services can be grouped together. Two land use services have been created, namely, built-up area and ecosystem service. Land use demand is calculated by assuming percentage increase and/or decrease as shown in Table 1.

This study applied the ecosystem service value (ESV) per unit area for seven land use types based on the equivalent coefficient value of nine ecosystem service functions (ESV_f): gas regulation, climate regulation, water supply, soil formation, waste treatment, biodiversity protection, food production, raw material, and recreation and culture which is proposed by [2] as shown in Table 2

Ecosystem service value (ESV) of seven land use types, namely, cultivated land, forestland, grassland, wetland, unused land, construction land, and water body, is calculated by multiplying total ESV value with area of each projected land use for BAU and ES scenarios in 2032 and compared with those of 2017 as the following equations:

$$ESV_k = \sum_f A_k \times VC_{kf} \tag{1}$$

$$ESV_f = \sum_k A_k \times VC_{kf} \tag{2}$$

$$ESV = \sum_k \sum_f A_k \times VC_{kf} \tag{3}$$

where ESV_k, ESV_f, and ESV are the value of the ecosystem services of land use type k, the value of ecosystem service function type f, and the total ecosystem service value, respectively. A_k is the area (ha) for land use type k, and VC_{kf} is the value coefficient (USD/ha/year) for land use type k and ecosystem service function type f.

Table 1 Percentage increase in land demand based on different scenarios of land use

Scenarios	Business as usual (BAU) (%)	Ecosystem services (ES) (%)
Built-up area	2	2
Ecosystem services	0.25	1.5

Table 2 Ecosystem service value of each land use type

Ecosystem service function (ESV$_f$)	Ecosystem service value (USD/ha/year)						
	Cultivated	Forest	Grass	Wetland	Unused	Construction	Water
Total ESV$_f$	1032.3	1948.9	809	9368.7	96.4	12.7	6873.8

Source: Mamut et al. (2018)

3   Results and Discussion

3.1   Land Use Change Between 2017 and 2032 for BAU Scenario

Land use projection in 2032 for BAU scenario (Fig. 1) reveals an increase in forest-land, construction land, and urban land, respectively, while wetland and unused land are decreased significantly. Ecosystem service value of wetland, cultivated land, and unused land is decreased in contrast with those of forestland, grassland, and construction. However, ecosystem service value of Huai Luang land use in 2032 tends to be decreased by 9.7×10^6 USD compared to ecosystem service value in 2017 as shown in Table 3.

3.2   Land Use Change Between 2017 and 2032 for ES Scenario

As shown in Fig. 2, forest, wetland, and construction land in Huai Luang river basin tend to be increased by 2032 for ES scenario compared to those of 2017, while cultivated land, grassland, and unused land have declined over time. Urban growth is accelerating a decrease in farmland and unused land. Forests and wetlands provide a wide range of ecosystem services as shown in Table 4. As a result, the total ecosystem service value from 2017 to 2032 is expected to be increased by 1.2×106 USD for ES scenario.

Land use projection in Huai Luang river basin in 2032 for both BAU and ES scenarios reveals a fact that biophysical and socioeconomic factors appear to be determinants of land use change at different development pathways. The increase in urbanization is expected as occurred before 2017, resulting to land without perceived benefit, i.e., unused land and wetland, tends to be declined by 2032. Regarding ES scenario, urban expansion rate is lower than those of BAU scenario, while wetland appeared to be much more increased by comparing with BAU due to its capability of providing higher ecosystem service value, i.e., climate regulation, water supply, and waste treatment. Urban growth will continue to increase by 2032 for both BAU and ES scenarios in accordance with government policy that aims to promote Udon Thani province as one of the central hubs in northeastern region. However, construction land gives no value of ecosystem services except recreation and culture value.

Total ecosystem service value of ES scenario is higher than those of BAU in 2032, 52.8×106 USD and 41.9×106 USD, respectively. It can imply that land use

Fig. 1 Land use change between 2017 and 2032 under BAU scenario

Table 3 Ecosystem service value change between 2017 and 2032 for BAU scenario

Land use	Ecosystem service value (USD) 2017	Ecosystem service value (USD) 2032	Ecosystem service value change (USD)
Cultivated land	193,989,816	193,775,097.6	−214,718.4
Forestland	78,926,704	87,745,929	8,819,225
Grassland	9,172,758	15,512,182	6,339,424
Wetland	124,866,034	12,291,734.4	−112,574,299
Unused land	214,171.2	0	−214,171
Construction land	380,796.8	407,009.6	26,212.8
Water body	109,319,325	109,429,304	109,979.2
Total	516,869,604.8	419,161,257	−97,708,347.8

Fig. 2 Land use change between 2017 and 2032 under ES scenario

Table 4 Ecosystem service value change between 2017 and 2032 for ES scenario

Land use	Ecosystem service value (USD) 2017	Ecosystem service value (USD) 2032	Ecosystem service value change (USD)
Cultivated land	193,989,816	190,769,040	−3,220,776
Forestland	78,926,704	85,974,288	7,047,584
Grassland	9,172,758	8,383,565	−789,194
Wetland	124,866,034	134,009,885	9,143,851.2
Unused land	214,171.2	0	−214,171
Construction land	380,796.8	402,539.2	21,742.4
Water body	109,319,325	109,429,304	109,979.2
Total	**516,869,604.8**	**528,968,621**	**12,099,016.2**

planning in accordance with ES scenarios will provide higher ecosystem service value in terms of climate regulation, water supply, and waste treatment from the increase in wetland area. Land use planner or policy makers and stakeholders in Huai Luang river basin can consider ES development pathways which can potentially contribute to ecosystem services and sustainable development goals and put it into practices, i.e., river basin and provincial plans.

4 Discussion

Scenario-based assessment of land use change in Huai Luang river basin, Udon Thani province, has been conducted by reviewing relevant information and categorized into two scenarios: business as usual (BAU) and ecosystem services (ES). Both scenarios are driven by six biophysical and socioeconomic driving forces in which BAU means land use change is continuing as it always do even though something unpleasant or unexpected has happened, while ES have focused on ecosystem and sustainability movement. As a result, ES scenario would be appropriate for environmentally friendly development than BAU scenario. However, the consideration on development pathways for land use planning depends not only on ecosystem perspective but also include socioeconomic prosperity. Land use planner and/or policy maker needs to be aware of mainstreaming sustainability concept into all future land use scenarios. Scenario-based assessment can help assess a wide range of alternative policy options. From results of study, forest and wetland provide higher value of ecosystem service than other land use, i.e., water supply, climate regulation, and soil formation. However, coordinating socioeconomic development and ecological protection is essential and must be addressed in the context of river basin management and provincial planning. All stakeholders should understand a key concept of ecosystem services to act wisely to achieve sustainability among economic prosperity, social welfare, and environment protection in the future.

Acknowledgments This work was supported by Strategic Wisdom and Research Institute, Srinakharinwirot University, Thailand.

References

1. Alcamo, J.: The SAS approach: combining qualitative and quantitative knowledge in environmental scenarios. In: Alcamo, J. (ed.) Environmental futures. The practice of environmental scenario analysis, vol. 2, pp. 123–150. Elsevier, Amsterdam (2008). https://doi.org/10.1016/S1574-101X(08)00406-7
2. Mamat, A., Halik, Ü., Rouzi, A.: Variations of ecosystem service value in response to land-use change in the Kashgar Region, Northwest China. Sustainability. **10**(1), 200 (2018). https://doi.org/10.3390/su10010200
3. Millennium Ecosystem Assessment: Ecosystems and Human Well-being Biodiversity Synthesis. World Resources Institute, Washington, DC (2005)
4. Owicki, D., Walz, U.: Gradient of land cover and ecosystem service supply capacities- a comparison of suburban and rural fringes of towns Dresden (Germany) and Poznan (Poland). Procedia Earth Planet Sci. **15**, 495–501 (2015). https://doi.org/10.1016/j.proeps.2015.08.057
5. Yirsaw, E., Wu, W., Temesgen, H., Bekele, B.: Effect of temporal land use/land cover changes on ecosystem services value in coastal area of China: the case of Su-Xi-Chang region. Appl Ecol Environ Res. **14**(3), 409–422 (2016). https://doi.org/10.15666/aeer/1403_409422
6. Mallampalli, V.R., Mavrommati, G., Thompson, J., Duveneck, M., Meyer, S., Ligmann-Zielinska, A., Druschke, C.G., Hychka, K., Kennt, M.A., Kok, K., Borsuk, M.E.: Methods for translating narrative scenarios into quantitative assessments of land use change. Environmental Modelling & Software. **82**, 7–20 (2016). https://doi.org/10.1016/j.envsoft.2016.04.011
7. Swart, R.J., Raskin, P., Robinson, J.: The problem of the future: sustainability science and scenario analysis. Global Environmental Change. **14**(2), 137–146 (2004). https://doi.org/10.1016/j.gloenvcha.2003.10.002
8. Welp, M., de la Vega-Leinert, A., Stoll-Kleemann, S., Jaeger, C.C.: Science-based stakeholder dialogues: theories and tools. Global Environment Change. **16**(2), 170–181 (2006). https://doi.org/10.1016/j.gloenvcha.2005.12.002
9. Houet, T., Aguejdad, R., Doukari, O., Battaia, G., Clarke, K.: Description and Validation of a "non path-dependent" model for projecting contrasting urban growth futures. Cybergeo: European Journal of Geography. **759** (2016). https://doi.org/10.4000/cybergeo.27397
10. Hasan, S., Shi, W., Zhu, X.: Impact of land use land cover changes on ecosystem service value – a case study of Guangdong, Hong Kong, and Macao in South China. PLoS ONE. **15**(4), e0231259 (2020). https://doi.org/10.1371/journal.pone.0231259

The Ability to Access Attractions for the Elderly Using Public Transport in Bangkok Metropolis

Sutatip Chavanavesskul and Pakorn Meksangsouy

Abstract The current population situation in Thailand has become an aging society. The elderly group is a new market for tourism business, which is normally targeted as high purchasing power group. This research aims to study the ability to access tourist attractions for the elderly by using public transport in Bangkok Metropolis. The study area provides many modes of public transport to choose, which are ease and convenience to travel. These will help the elderly in the city to live and to do activities outside their homes. The service area of public transport within a specific distance, under the limitation of the elderly, can walk within 400 m to reach public transport stations, and an average walking time is 1.21 m/s (4.34 km/h). The results show that appropriate tourist attractions for the elderly reached by public transport in Bangkok Metropolis consist of 403 places, which were found in the Bangkok inner zone. These comprise of 87 department stores, 237 tourist attractions, and 79 parks.

Keywords The elderly · Ability to access · Public transport · Bangkok Metropolis

1 Introduction

The current population situation in Thailand has been changing over decades. It is stepping into an aging society with a population aged over 60 years old, counting more than 10% of the total population and/or having a population aged 65 and over more than 7% of the total population [1]. Population projection in 2040 in Thailand shows that the population is living longer while the level of fertility and morality rate is declining. As a result, proportion of Thai elderly will increase to 31.28% from the present rate of 18% (2020) (12 million people) [2]. This causes a change

S. Chavanavesskul (✉) · P. Meksangsouy
Department of Geography, Faculty of Social Sciences, Srinakharinwirot University,
Bangkok, Thailand
e-mail: sutatip@g.swu.ac.th

© The Author(s), under exclusive license to Springer Nature
Switzerland AG 2023
W. Boonpook et al. (eds.), *Applied Geography and Geoinformatics for
Sustainable Development*, Springer Geography,
https://doi.org/10.1007/978-3-031-16217-6_19

247

in the population structure, which has a wide impact on the necessary infrastructure in the country and economic, social, and political systems, such as increased government budget management, lower employment, and decreased GDP. The elderly is neglected and may be abandoned. Therefore, based on these data, it was assigned as a national agenda to prepare for transition into an aging society, where the elderly people can adapt to the environment, enhance skills, live with dignity, enjoy life, and participate in society according to their physical and mental condition [3–6]. In addition, many elderly people can have an active lifestyle, are able to live smoothly and businesslike, can take care of themselves, and can be strong and able to travel and live outside by themselves. And so facilities in an urban area are developed for the elderly such as an elevator to help them go up and down comfortably and smooth pavement and ramps wherein they don't have to raise their legs, as well as an urban public transport for them to use. Therefore, it is important to facilitate the elderly living outside for them to be able to participate in social activities and open opportunities for social interaction. Hence, the elderly turned attention to tourism. It is a new market source in the tourism business [7]. In Thailand, the elderly group has higher purchasing power than Gen X and Millennials with 1.5 and three times purchasing power, respectively [8]. Thus there is a need for recreational tourism, relaxation, nature time, leisure, and easy access to tourist attractions. Especially in Bangkok Metropolis, there are many public transport modes to choose from, and it is easy and convenient to travel without finding parking and going to many places in a limited and urgent time as well as to facilitate the connection to various places by walking without wasting time on traffic.

However, due to the physical conditions of some elderly, there are limitations that need to be confronted such as walking long distance to access various tourist attractions. Therefore, facility preparation is necessary to be conducive and easily accessible to the elderly. Finding suitable tourist attractions for the elderly that are easy to walk and easy to connect to various public transport systems in Bangkok Metropolis will encourage the elderly in the city to go out and to do activities outside their homes, to feel relaxed, to have a clear mind, and to feel less depressed. The top two types of tourism that the elderly pay attention to are (1) cultural tourism (learning traditions, tracing the past, watching cultural performances) and (2) health and recreation tourism which relates to create an energetic, body relaxation and maintain the brain and mood stability, doing a variety of artworks, shopping in department stores [9]. Therefore, this research aims to study the ability to access tourist attractions for the elderly by public transport in Bangkok Metropolis as well as guidelines in order to prepare for the elderly society in Thailand.

2   Methodology

2.1   Research Scopes

2.1.1 Scope of the area: Bangkok Metropolis is selected as a study area. It has approximately 1569 km^2 (as shown in Fig. 1).

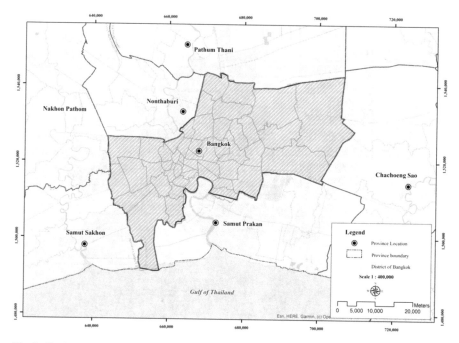

Fig. 1 Study area: Bangkok Metropolis

2.1.2 Scope of content: this research focuses on existing and future public transport systems and suitable tourist attractions for the elderly in Bangkok Metropolis.

2.1.3 Scope of time: this research collected data from October 2021 to April 2022.

2.2 Field Survey

The study of the ability to access tourist attractions for the elderly from public transport in Bangkok Metropolis consists of five steps. Details are shown as follows:

First, review documents on the elderly, tourist attractions, and seven modes of public transport in Bangkok Metropolis: sky train (BTS), underground train (MRT), expressway bus (BRT), public bus, railway, the Saen Saep boat, and the Chao Phraya boat.

Second, data collection by using Global Positioning System (GPS). Data consists of tourist areas around public transport and accessibility network to public transport stations, which have a total of 698 tourist attractions and 4512 public transport pickup points (shown in Table 1 and Figs. 2 and 3). Then, all data are transferred into geographic database in order to analyze further steps.

Third, preparation of spatial database by validating and importing data into GIS program, classification of tourist attractions for the elderly according to the nature

Table 1 Type of spatial data

Type	Detail	Number of stations
Public transport	1. Public bus station	4222
	2. BTS station	52
	3. MRT station	40
	4. BRT station	35
	5. Railway station	52
	6. Saen Saep pier	27
	7. Chao Phraya pier	84
Tourism attractions	1. Department stores	119
	2. Tourist attractions: museums, historic sites, natural attractions, historical/religious sites, and cultural arts	374
	3. Public parks	205

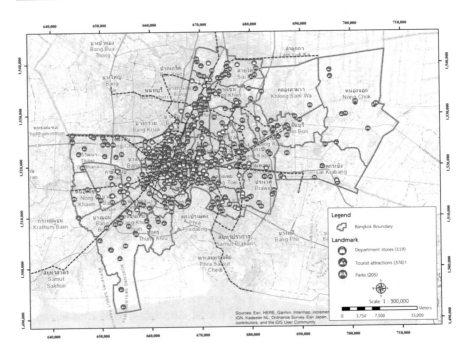

Fig. 2 Tourism attractions in Bangkok Metropolis

and environment [10–12]. It can be divided into three groups: department stores, tourist attractions, and parks. All spatial data are validated for accuracy in the same time.

Fourth, data analysis: network analysis is a main study approach in order to find service areas for the elderly to access from public transport stations to tourist attractions in Bangkok Metropolis [13–16]. Determine the usage of roads and service

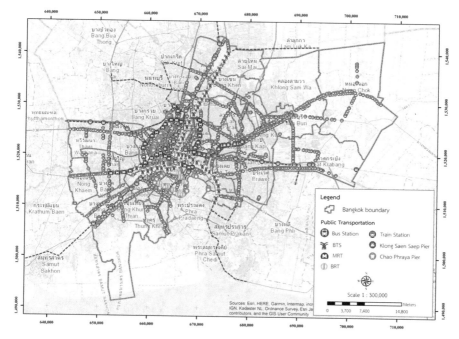

Fig. 3 Public transport station in Bangkok Metropolis

areas depending on distance and travel times. Then, service radius from seven types of public transport stations is calculated, which are controlled by time limitation of actual road travel time. In order to find tourist attractions within the service area of public transport within a specific distance, the 400-m radius is assigned under condition of walking ability of the elderly within 400 m from pickup public transport stations (and walking time is 1.21 m/s or 4.34 km/h) [17]. After that, the service areas from seven types of public transport stations are analyzed in order to explore the appropriate tourist attractions in the service area within 400 m (as shown in Fig. 4).

Fifth, analysis of appropriate tourist attractions within the service area of 400 m from seven types of public transport stations. Then, these service areas are represented into map.

3 Findings

There are 615 tourist attractions found for the elderly to access using public transport in Bangkok Metropolis, which are calculated from the 400-m radius of the public transport stations. Table **2** presents appropriate tourist attractions within the service areas. It shows that the service areas cover 364 bus stations, 60 BTS stations,

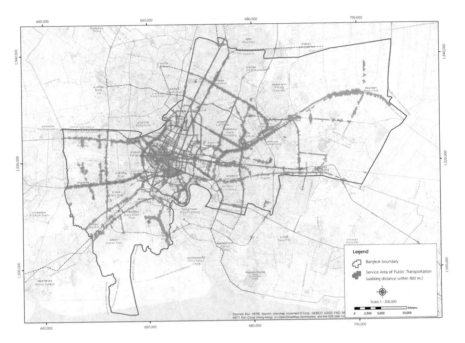

Fig. 4 Service areas in Bangkok Metropolis

Table 2 Appropriate tourism attractions within the service areas

| Type of tourism attractions | Number of service areas within 400 m from station | | | | | | | Total |
	Bus	BTS	MRT	BRT	Railway	Saen Saep	Chao Phraya	
Department store	80	24	17	2	2	8	6	**139**
Tourist attractions	218	24	29	3	4	8	87	**373**
Parks	66	12	8	1	2	2	12	**103**
Total	**364**	**60**	**54**	**6**	**8**	**18**	**105**	**615**

54 MRT stations, six BRT stations, eight railway stations, 18 Saen Saep piers, and 105 Chao Phraya piers (as shown in Figs. 5, 6, 7, 8, 9, 10, and 11).

Simultaneously, the research also explores the ability to access attractions for the elderly from all of the public transport stations in Bangkok Metropolis by applying overlay method. It shows the service area from all the public transport stations within 400 meters. It consists of 403 places, which are divided into 87 department stores, 237 tourist attractions, and 79 parks (as shown in Fig. 12).

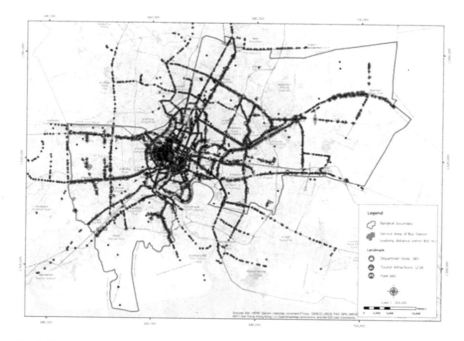

Fig. 5 Tourism attractions in the service area from bus stations

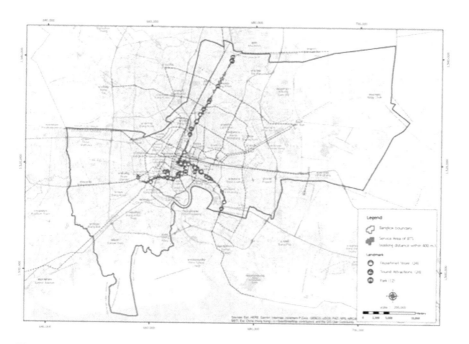

Fig. 6 Tourism attractions in the service area from BTS stations

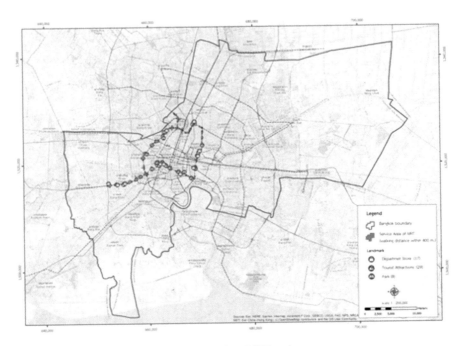

Fig. 7 Tourism attractions in the service area from MRT stations

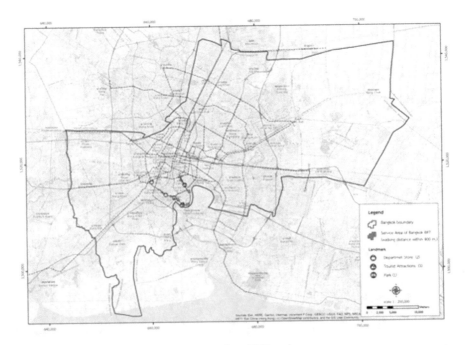

Fig. 8 Tourism attractions in the service area from BRT stations

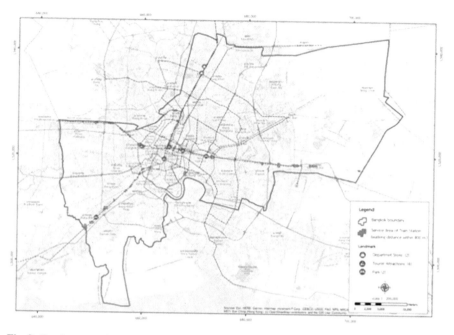

Fig. 9 Tourism attraction in the service area from railway stations

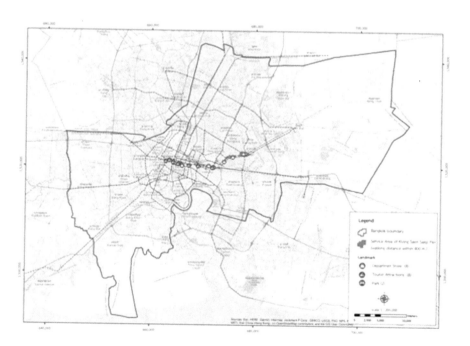

Fig. 10 Tourism attraction in the service area from Saen Saep piers

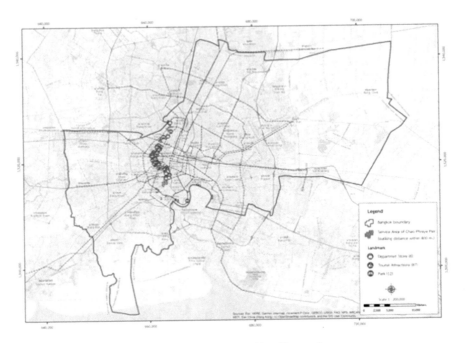

Fig. 11 Tourism attractions in the service area from Chao Phraya piers

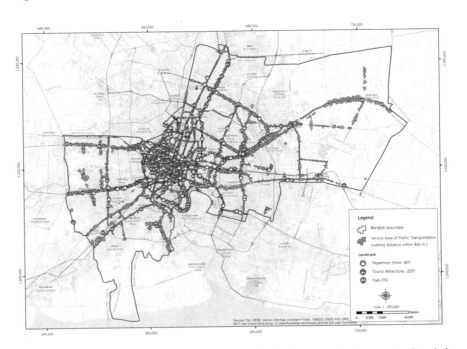

Fig. 12 The ability to access attractions for the elderly from public transport in Bangkok Metropolis

4 Discussion and Conclusion

4.1 The Ability to Access Tourist Attractions for the Elderly by Public Transport in Bangkok Metropolis

The most travel methods for the elderly in Bangkok Metropolis when they go outside from home are on foot and public transport. They often travel alone or with friends or partner. They also decide to visit tourist attractions that are safe and comfortable, have seats and foods, and are easy to access. Normally, the elderly walks for a short distance, in low speed walking, or spends a short time walking from the public transport stations to tourist attractions [18, 19]. Thereby, walking is an important part of connecting public transport networks to access tourist attractions that are located in urban areas. There are 403 attraction places for the elderly from all the public transport stations in the service area. They are found in the inner zone of Bangkok Metropolis, where it represents as a cluster pattern of tourist attractions such as museums, natural attractions, historical/religious sites, and cultural arts, namely, Bang Khun Phrom Palace, The Royal Navy Dockyard Museum in Honor of His 84th Birthday, and the National Museum. On the other hand, some attractions are located in the middle and outer zones of Bangkok Metropolis such as Robinsons, Seacon Square department store, and Rama 9 Chalermprakiet park. However, no tourist attractions for the elderly are found in some service areas.

4.2 A Guideline to Prepare for the Elderly Society of Thailand

Bangkok Metropolis has often seen problems with damaged pedestrians. The distance from public transport stations to tourist attractions is long, and the environment does not support walking, especially for the elderly who has physical limitation and difficulty with tourism. Traveling for the elderly is an activity that requires advance planning like making decision for tourist places, how to travel, route selection, travel duration, amenities, or activities during tourism. These are all factors that the elderly use to decide on tourist attractions. In the same time, Thailand is now confronted with an increasing number of aging group. Therefore, tourist attractions should be planned in order to develop tourism destinations for elderly tourism. For example, proper planning many environment following universal design concept for the elderly, pedestrian and walkway improvement for public transport stations, designing ramps at connecting points and intersections in order to support commuters, providing rails and handrails at stairs, using non-slip pavement materials around tourism places, convenient, improvement and development at tourist attractions in order to attract and facilitate activities for the elderly.

Acknowledgments This research has been granted research funding by the Faculty of Social Sciences, Srinakharinwirot University, Bangkok, Thailand. Researchers would like to convey utmost gratitude for the funding.

References

1. United Nations Department of Economic and Social Affairs, Population Division: World Population Ageing 2007, pp. 470–471, New York (2007)
2. National Statistical Office: Health and Welfare Survey 2019. Bangkok (2019)
3. Foundation of Thai Gerontology Research and Development Institute. (November, 2021). In future 20 years, NESDB expects the number of Thailand elderly jump to 31 percent. Homepage, https://thaitgri.org/?p=39327, Last accessed 15 Feb 2021
4. Sangkhakon, K., Bunyanupong, B., Thiensiri, C.: Assessing the potential of Slow Tourism attractions in the Upper North suitable for elderly tourists. Social Research Institute Chiang Mai University Chiang Mai University, Chiang Mai (2012)
5. Gray, R., Pattaravanich, U., Jamchan, C., Suwannoppakao, R.: The concept of the definition of the elderly: social psychology and health perspectives = New Concept of Older Persons: The psycho-social and health perspective. Research Institute Population and Society, Mahidol University. Homepage, http://www.ipsr.mahidol.ac.th/ipsrbeta/FileUpload/PDF/Report-File-419.pdf. Last accessed 15 Feb 2021
6. Department of Tourism. Thai tourism standard service for the disabilities, the elderly and families with children and pregnant women. (Publication). Ministry of Tourism & Sports (2012)
7. Ratanaphaitoonchai, J.: Elderly tourist market: new opportunities for Thailand growth to receive AEC. Krungthep Turakij Newspaper Section: ASEAN: 1 (2014)
8. Prasongthan, S., Thanadornsum, K., Charoenbunprasert, S.: Thai senior tourists: explore to travel constraints, recreational activities, and travel intention. Rajamangala University of Technology Tawan-ok Social Science Journal. 10(1), 119–131. Chonburi (2021)
9. De Aguiar Eusebio, M.C., Aibeo Carneiro, M.J., Kastenholz, E., Dourado Alvelos, H.M.: Potential benefits of the development on an european programme of social tourism for seniors. Homepage, http://www.ecalypso.eu/webs/steep/web/documentos/DOC_2_4.pdf. Last accessed 15 Feb 2021
10. World Tourism Organization. UNWTO Tourism Definition. Madrid (2019). https://doi.org/10.18111/9789284420858. Last accessed 15 Feb
11. Tourism Authority of Thailand. Homepage, https://www.tat.or.th/th. Last accessed 15 Feb 2021
12. Boonrawd, C.: The needs of the elderly in Bangkok for tourism business. In thesis Master's degree in Management, College of management, Mahidol University, Bangkok (2014)
13. Meyer, D., Miller, E.: Urban Transportation Planning: a decision-oriented approach. McGraw Hill, New York (1984)
14. Rujankanoknat, J.: Principles of transportation planning. Master of Engineering in thesis Master's degree in Construction management and Utilities, Department of Civil Engineering, Faculty of Engineering Chulalongkorn University. Bangkok (2008)
15. Chavanavessakul, C.: The study on the availability of public transport services in Bangkok. Journal of Social Sciences. 14, 84–91. Srinakharinwirot University (2011)
16. Khamnamul, N.: Technology of urban public transport: public transport in Bangkok, Thailand. Institute of Scientific and Technological Research (TISTR)/Seven Printing Group Co., Ltd., Bangkok (2004)
17. Alves, F., Cruz, S., Ribeiro, A., Silva, A.B., Martins, J., Cunha, I.: Walkability index for elderly health: a proposal. Sustainability. 12(7360), 1–27 (2020)
18. Samphantharak, Petcharyiim T, et al. Happiness through age: a built environment that is friendly to elderly people. E.T. Bangkok (2017)
19. Ngampraphasom, P.: A study of potential and preparation for tourism management, community for the elderly with participation: a case study of Ban Mo Luang tourism tommunity Mae Mo District, Lampang Province. Journal of Humanities and Social Sciences. 8(2), 177–190 (2017)

Index

Printed in the United States
by Baker & Taylor Publisher Services